Bibliotheca Diatomologica   Volume 68

Small marine Achnanthales (Bacillariophyceae) from coral reefs off Polynesia (South Pacific)

Specificities and biogeography

Catherine Riaux-Gobin, Andrzej Witkowski, Richard W. Jordan and Michel Coste

With 30 figures and 6 tables

J. Cramer

in Borntraeger Science Publishers · Stuttgart · 2023

Editors   Prof. Dr. Dr. h.c. H. Lange-Bertalot, Frankfurt
Prof. Dr. J. P. Kociolek, Boulder

Author's addresses

*Riaux-Gobin, C., CRIOBE-UAR 3278 CNRS-IRCP-UPVD, Perpignan, France

Witkowski, A., University of Szczecin, Institute of Marine and Environmental Sciences, Szczecin, Poland
(andrzej.witkowski@usz.edu.pl, https://orcid.org/0000-0002-0442-4586)

Jordan, R.W., Faculty of Science, Yamagata University, Yamagata, Japan
(sh081@kdw.kj.yamagata-u.ac.jp, https://orcid.org/0000-0002-8997-7349)

Coste, M., French National Institute for Agriculture, Food, and Environment-INRAE (EABX), France
(michelcoste24@gmail.com, https://orcid.org/0000-0003-2863-0365)

*Corresponding author:
catherine.gobin@univ-perp.fr, https://orcid.org/0000-0002-6128-8947

This volume was edited by Bart Van de Vijver

Cover: Napuka lagoon land scape, with SEM illustration of *Cocconeis napukensis*.

We would be pleased to receive your comments on the content of this book: editors@schweizerbart.de

ISBN 978-3-443-57059-0
ISSN 1436-7270 (Bibliotheca Diatomologica)

Information on this title: **www.borntraeger-cramer.de/9783443570590**

© 2023 Gebr. Borntraeger Verlagsbuchhandlung, Stuttgart

All rights reserved including translation into foreign languages. This book or parts thereof may not be reproduced, digitalized or stored in any form without permission of the publishers.

Publisher: Gebr. Borntraeger Verlagsbuchhandlung
　　　　　Johannesstr. 3A
　　　　　70176 Stuttgart, Germany
　　　　　mail@borntraeger-cramer.de
　　　　　www.borntraeger-cramer.de

∞ Printed on permanent paper conforming to ISO 9706-1994
Printed in Germany

Content

1.	Abstract...	4
2.	Introduction..	4
3.	Material and methods	8
4.	Results...	10
4.1.	Marine Achnanthales from French Polynesia (2010–2020 survey)..	10
4.2.	Selected Achnanthales recently described from Polynesia	16
4.3.	New details on two pantropical taxa	24
4.4.	Rare and unnamed tropical taxa..	24
4.5.	Morphotypes or formae...	25
4.6.	Ill-defined taxa	26
4.6.1.	*Cocconeis placentula* Complex	26
4.6.2.	*Cocconeis scutellum* Complex..	28
4.6.3.	Loculatae section..	29
5.	Discussion	30
5.1.	South Pacific assemblages and historical records	30
5.2.	Assemblages from other tropical basins..................................	34
5.2.1.	Tropical Indian Ocean..	34
5.2.2.	West Atlantic (Florida and Caribbean Arc)................................	35
5.3.	Marine eukaryote endemism, myth or reality............................	40
5.4.	Role of insularity, position in the SEC and geologic past...........	41
5.5.	Concluding remarks	44
6.	References	45
	Figures 4–30...	55

1. Abstract

A 2010–2020 survey of small-sized marine Achnanthales (Bacillariophyceae) undertaken in Central Polynesia (South Pacific) was largely focused on ultrastructure using the scanning electron microscope (SEM). The degree of colonization, species richness and emergence of new diatom taxa appeared to vary according to the geologic past, presence of coral reefs and degree of insularity (insulation) of each island. The Tuamotu atolls, active coral structures formed during the ancient geologic past, are characterized by low diatom colonization, with a relatively high species richness. High volcanic islands (younger geologic structures with more or less restricted coral reefs, e.g. Society and Austral Archipelagos) are enriched with numerous pantropical and cosmopolitan taxa, and a variable number of newly described taxa. Among the Marquesas Islands, characterized by flooded fossil reefs and water masses enriched with particulate matter, Nuku Hiva was found to be poorly colonized, with a low Achnanthales diversity. Several recently published taxa from French Polynesia are here presented, with in part unpublished observations and new illustrations. Several taxa first described from the Indian Ocean are also present in French Polynesia and can be considered as pantropical, whereas some of the others are presently listed from only one oceanic basin. A Venn diagram permitted the first comparison between assemblages studied with the same methodologies from French Polynesia (Central Polynesia), New Caledonia (Melanesia) and Mascarenes (Indian Ocean). The assemblages from the tropical West Atlantic are briefly discussed. Some taxa with slight morphological differences from their original description may signal morphotypes or *formae* as possible 'cryptic taxa'. 'Potential endemism' is briefly discussed with 11 taxa pertaining to the genera *Amphicocconeis*, *Astartiella*, *Cocconeis* and *Xenocconeis*, yet only listed from the Pacific, and here briefly described. In addition, new details are added to the description of some rare pantropical taxa.

2. Introduction

Coral reef environments are generally recognized as highly diversified habitats through their high turn-over and complex food webs (Goreau et al. 1979, Pandolfi et al. 2003). The diatom floras are particularly diversified in tropical environments, especially among benthic diatoms (i.e. Riaux-Gobin et al. 2011b, Lobban et al. 2012). A survey from the Florida Keys (Tropical West-Atlantic) by Montgomery (1978) exemplifies this with SEM illustrations of numerous Achnanthales (ca. 60) demonstrating an impressive marine diatom diversity in tropical environments.

The Pacific Ocean is composed of three major 'cultural areas': Melanesia, Micronesia and Polynesia (Fig. 1), with Easter Island located in the most southeastern part of Polynesia and islands such as the Galápagos in the most eastward part of the Pacific Basin. Diatom monographs mainly concerned 1) Micronesia (Central West Pacific, i.e. Guam & Yap Islands and Moen Island: Navarro & Lobban 2009, Lobban et al. 2012, Lobban 2015; Park et al. 2018), 2) Polynesia

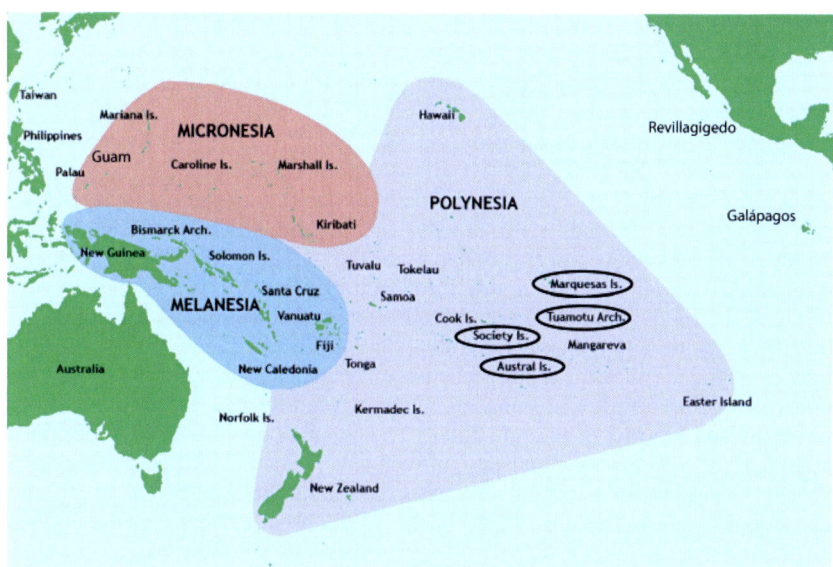

Fig. 1. Pacific Ocean culture areas: Melanesia (South-West Pacific), Polynesia (Central Pacific) and Micronesia (North-West Pacific). With position of Galápagos (Equatorial Eastern Pacific) and Revillagigedo (Mexico). Modified from: https://upload.wikimedia.org/wikipedia/commons/thumb/a/aa/Pacific_Culture_Areas.png/800px-Pacific_Culture_Areas.png; no copyrights.

(Central Polynesia-French Polynesia: Ricard 1975, Ricard 1977, Coste & Ricard 1990, Riaux-Gobin et al. 2021c; Easter Island: Navarro 2002; South Polynesia: New Zealand, from temperate to subtropical: Foged 1979), 3) Melanesia (Fiji Island: Foged 1987; New Caledonia, freshwater diatoms: e.g. Maillard 1978, Le Cohu 1985, Moser et al. 1998, and marine: Witkowski 1998 *in* Moser et al. 1998, Riaux-Gobin et al. 2022), and 4) Equatorial East Pacific (i.e. Galápagos Island: Hendey 1971, Stidolph et al. 2012, Łopato 2016; Revillagigedo Island: Siqueiros Beltrones et al. 2021). This relative scarcity of marine studies in such a large area (Figs 1–2) offers a wide range for further investigations, particularly concerning small and rare taxa.

French Polynesia extends over ca. 4167 km^2 (Fig. 2), with five archipelagos originating from several distinct hot spots lasting from 50 My BP (My BP = million years before present; e.g. atolls from the north-west Tuamotu Archipelago) to sub-actual (i.e. 300 000 y BP, Tahiti-Iti from the Society Archipelago – the smaller Tahiti volcano), even to actual (e.g. Mehetia Island, an active Society volcano). Depending on their age and geology, some islands are composed of volcanic high structures surrounded by more or less extended coral reefs, whereas older structures evolved into atolls, such as the Tuamotu Islands. Depending on complex geological events, some atolls are also present on younger archipelagos, such as the Gambier. All these atolls are doomed over the course of geological time to become drowned guyots.

A special focus can be shed on the Marquesas Archipelago [geologic age from 5.5–4.9 My BP to 2.5–1.2 My BP along a north-west to south-east alignment (cf. Chauvel et al. 2012, fig. 1), with an intricate origin for these islands and also for the Marquesas Plateau itself]. The high Marquesas Islands (e.g. Nuku Hiva) experienced deep flooded coral reefs, ca. 80–95 m deep (Rougerie et al. 1992). Some other coral structures from this archipelago are less deeply flooded, such as several submerged atolls (e.g. Lawson Bank with a coral reef more or less 5m deep), and quasi-atolls (e.g. Fatu'Uku). This contrasted evolution of islands within the Marquesas gave rise to various hypotheses (see arguments about complex isostatic movements, according to assumptions by S. Jourdan* and Bouniot & Jourdan 2009). The complex geology of Marquesas, in conjunction with sub-actual eustatic movements [sea level rise due to deglaciation, around 14 000 y BP, with water mass cooling unfavorable to coral growth, particularly in the north of the Pacific Basin (Guille et al. 2002, Legendre 2003)], resulted in the above cited contrasted geomorphology among islands.

Rapa Island, a high volcanic island from Austral, has no associated coral reef, while the geologic past and hydrologic conditions of this island do not seem incompatible with their establishment (Montagioni 2015). The latter remark highlights the complexity of the geologic events in the South Pacific and their implication on coral reef building.

The complex geologic past (age and origin) of the South Pacific Islands, particularly from French Polynesia (Brousse et al. 1990, Diraison 1991, Le Dez et al. 1996, Guille et al. 2002), lead to contrasted geomorphology from one archipelago to another, or even from one island to another. For example, Tetiaroa as well as Tupai (Society) are atolls while other islands of this archipelago are high islands or semi-atolls, with Mehetia being an active volcano with, up to now, no associated reef.

Benthic diatoms are eukaryotes e.g. micrometric primary producers with a siliceous frustule, requiring a substratum to establish colonies. Their behavior can be driven by the chemical and physical environment: silica concentrations, photic zone light levels, nutrients, salinity, suitable hydrodynamic exposure and substrate, insularity. Thus, it is interesting to think about the existence of these micro-organisms as a possible function of their environment linked to the geologic past and degree of insulation of each island. The South Equatorial Current (SEC) with a westward drive (Rougerie et al. 1997, Martinez et al. 2009) may prevent isolated islands from directly influencing other islands, thereby reinforcing their geographic isolation.

---

*see https://kn0l.wordpress.com/nos-collections/collection9-geographie/les-atolls-et-presquatolls-des-marquises-un-pave-dans-le-lagon/

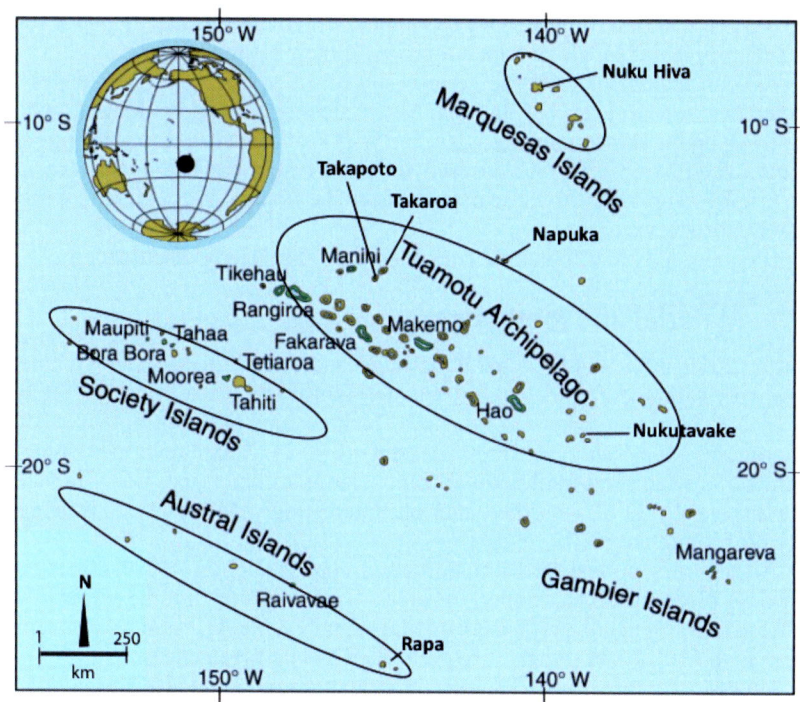

Fig. 2. French Polynesia position, with Marquesas, Tuamotu, Society, Gambier and Austral Archipelagos, and position of each sampled island during the 2010–2020 survey: Nuku Hiva, Takaroa, Takapoto, Napuka, Nukutavake, Tahiti, Moorea, Tetiaroa, Raivavae, Rapa. Modified version of:
https://www.google.com/search?client=firefox-b-d&q=41467_2017_2695_Fig1_HTML+; no copyrights.

The CORDIA project (COral Reef DIAtoms, Labex, 2010–2012) was focused on the diatom diversity from Society and Tuamotu. One goal of CORDIA was to study the order Achnanthales Silva 1962 through a geomorphologic reef gradient (from high volcanic islands of the Society, to atolls from Tuamotu). Further samplings (2013–2020) permitted the investigation of several other South Pacific islands, from Marquesas to Austral, and led to the description of several new taxa.

Eleven small and marine Achnanthaceae, to our knowledge only present in Polynesia, are herein illustrated using the SEM, with comments on their similarities to other taxa. A short description is given for each taxon, with some new illustrations (however, if necessary, it is recommended that the reader consults the full original description). New morphological details are also given for some rare pantropical taxa, along with a revision of their biogeography.

A Venn diagram allows the comparison of the taxonomic assemblages reported herein from French Polynesia, New Caledonia and Mascarenes (data obtained using the same methodologies, and with a similar effort -concerning sampling and SEM observation- on each sector). Finally, the Pacific assemblages are briefly compared with those detailed from the tropical West Atlantic through historical surveys.

Biogeography and 'potential endemism' are also briefly discussed.

3. Material and methods

During 2010–2020, ten French Polynesian islands, pertaining to four Archipelagos, were sampled: Tahiti-Nui – the larger Tahiti volcano –, Moorea, Tetiaroa (Society); Napuka, Takaroa, Takapoto, Nukutavake (Tuamotu); Nuku Hiva (Marquesas) and Raivavae, Rapa (Austral) (Fig. 2). Mangareva, pertaining to Gambier, a wide semi-atoll with volcanic islands still merging, was sampled in late 2021 and will be presented later on. Interestingly, Mangareva has an easterly position in the SEC.

All samples collection from Polynesia and Melanesia (2010–2020) took place in September–December, whereas the samplings in the Mascarenes (2005–2009) were all performed in May–June (Table 1). As a consequence, these Indo-Pacific samplings were performed during the dry season.

Diverse habitats, mostly marine intertidal, were sampled on each island, representing in total ca. 150 marine samples, which were observed during 240 SEM sessions. The investigated substrates were diverse, from mineral to plant and animal: coral sands, dead corals, intertidal marine sediments (i.e. surficial muds from mangroves), diverse macroalgae and short turfs, but also invertebrate scrapings, such as scrapings of *Holothuria atra* Jaeger and bivalve mollusks, particularly *Pinctada margaritifera* Linneaus. Note that the sampled mangroves from French Polynesia (e.g. from Moorea Island) are not natural units, but settlements of human origin (*Rhizophora stylosa* Griffith was introduced in Moorea in 1930–1935; Cavaloc 1988). Scrapings of numerous marine turtles [ca. 120 specimens pertaining to four species, from French Polynesia, Martinique (Caribbean arc) and Guyana] were also studied, in collaboration with Damien Chevallier (turtle ethology, Institut Pluridisciplinaire Hubert Curien, UMR 7178, CNRS-Unistra) as part of his ANTIDOT project. The latter studies mainly focused on gomphonemoid genera and possible 'commensalism' (Riaux-Gobin et al. 2017a,b; Riaux-Gobin et al. 2021a).

Raw material (RM) was preserved with formalin (10% final dilution) or absolute ethanol. Due to the small size of the taxa, the descriptions of new monoraphids were based on very fine structures, e.g. the valvocopulae and areola hymenate structure, which are difficult or impossible to observe in the light microscope (LM). The scanning electron microscope (SEM) was systematically used and commented here (LM illustrations of several taxa are available in the

Table 1. Sampling dates (Indian Ocean and South Pacific Ocean). See corresponding taxonomic lists *in* Table 2.

Réunion	Indian Ocean	02–23 June 2005
Réunion & Rodrigues	Indian Ocean	13–18 June 2007
Réunion	Indian Ocean	15–18 May 2009
Moorea	Society	14–21 Oct. 2010
Tahiti-Nui	Society	23–26 Oct. 2010
Moorea	Society	16–17 Sept. 2012
Napuka	Tuamotu	23–29 Sept. 2012
Moorea	Society	04–06 Oct. 2012
Tetiaroa	Society	09–10 Oct. 2012
Moorea	Society	11–16 Oct. 2012
Tahiti-Nui	Society	18–22 Oct. 2012
Moorea	Society	24 Oct.–09 Nov. 2012
Moorea	Society	03 Oct. 2013
Takapoto	Tuamotu	05–09 Oct. 2013
Takaroa	Tuamotu	11–15 Oct. 2013
Tahiti-Nui	Society	18 Oct. 2013
Nuku Hiva	Marquesas	21 Nov.–05 Dec. 2015
Moorea	Society	06–11 Dec. 2015
Nukutavake*	Tuamotu	Oct. 2016
Raivavae	Austral	05–09 Oct. 2018
Moorea	Society	15–16 Oct. 2018
New Caledonia	Melanesia (South Pacific)	04–10 Nov. 2019

*Sampled by V. Paravicini

original descriptions). Every SEM stub was prepared with two drops of raw material filtered onto a Whatman® Nuclepore filter (1 μm pore size, 13 mm in diam.) and rinsed twice with deionized water (Milli-Q®) to remove salts. Filters were air-dried, mounted onto aluminum stubs, before coating with gold-palladium alloy (EMSCOP SC 500 sputter coater) and examined with a Hitachi FEG S4500 model I SEM, operated at 5 kV, calibrated with a Silicon grating TGX01 (C2M, Perpignan, France). Diatoms (often <10–20 μm in length) were determined at high magnification. Diatoms were roughly classified within abundance categories (Table 2), after examination of ca. 1/8 of the filter at high magnification. When the cell density was important, the evaluation of abundance was done on a fraction of the transect. The latter rough index makes it possible to assess the rarity of each taxon. SEM observation effort was more or less the same for the Mascarenes and the French Polynesian assemblages, first with a Hitachi S520 (W tungsten filament source) and afterwards with a Hitachi FEG S4500 model I SEM (cold-tip field-effect source). The presence-absence of rare taxa from both oceanic basins can thus be compared.

GPS location, water temperature and salinity were recorded. Nutrients and heavy metals were punctually recorded, but are not commented on here.

Typography of authorities for each taxon is intentionally simplified in eluding the first names (full typography available in the original descriptions). Acronyms are used hereafter, such as SV (sternum valve), RV (raphe valve), SVVC (sternum valve valvocopula) and RVVC (raphe valve valvocopula).

The R software (R Core Team 2018) vegan package 'Venn diagram' was used to compare the assemblage lists, and evaluate the degree of similarity between them (Venn 1880).

4. Results

4.1. Marine Achnanthales from French Polynesia (2010–2020 survey)

In this study, ca. 73 small-sized marine Achnanthales were recorded during the 2010–2020 survey in French Polynesia (Table 2). Typical freshwater taxa are not included here. The taxa pertain to eleven genera (*Achnanthes* Bory 1822, *Achnanthidium* Kützing 1844, *Amphicocconeis* De Stefano & Marino 2003, *Anorthoneis* Grunow 1868, *Astartiella* Witkowski, Lange-Bertalot & Metzeltin *in* Moser 1998, *Cocconeis* Ehrenberg 1836, *Madinithidium* Desrosiers, Witkowski & Riaux-Gobin (Desrosiers et al. 2014), *Planothidium* Round & Bukhtiyarova 1996, *Pseudachnanthidium* Riaux-Gobin (Riaux-Gobin & Witkowski 2015d), *Vikingea* Witkowski, Lange-Bertalot & Metzeltin 2000, *Xenococconeis* Riaux-Gobin (Riaux-Gobin et al. 2014b), *Upsilocoсconeis* Riaux-Gobin, Witkowski & Risjani (Riaux-Gobin et al. 2022), and *Navithidium* Al-Handal & Romero (Al-Handal et al. 2021). Several rare taxa (often only documented by their SV) are not mentioned in this study, among which are several *Planothidium* and *Achnanthes* taxa that need more investigations before formal classification.

Cocconeis was dominant (Table 2) comprising ca. 49 taxa, most of which were found as epiphytes on short turfs and large macroalgae. Among the latter *Cocconeis*, twelve taxa (published or in press, Table 2, asterisk) are only reported from Polynesia, and eleven of them are briefly described and illustrated below, with some additional unpublished iconography (see '*Selected Achnanthales recently described from Polynesia*'). The island from which each new species was first discovered is indicated, as well as its similarities with closely-related species. Table 6 gives the location of each type material.

Besides *Cocconeis*, several other genera, such as *Amphicocconeis*, *Astartiella* and *Madinithidium*, were also diverse (Table 2), as well as *Planothidium* and *Achnanthes* that were underestimated in the present study (due to the lack

Table 2. List of marine Achnanthales from French Polynesia (2010–2020 survey), New Caledonia (NC30–NC2/2019 survey, 2022b) and Mascarenes (2005–2009 survey, Riaux-Gobin et al. 2011b). All surveys conducted with the same methodology. Rough abundance based on the number of frustules-valves present in ca. 1/8 of Millipore filter (13 mm in diameter, 1 μm pore size), two drops of raw material.

A = abundant (>1000); M = moderately abundant (ca. 500); R = rare (10–50); VR = very rare (1–5); – = absent
* = illustrated in the present study
** = *Navithidium delicatissima* (Simonsen) Al-Handal, Romero & Wulff
*** = *Upsiloconeis dapalistriata* (Riaux-Gobin, Romero, Compère & Al-Handal) Riaux-Gobin, Witkowski & Risjani comb. nov. (2022)
Present on Scattered Islands (1), Madagascar (2) (from other suveys, same team-methodology)
# = unique to French Polynesia, see Venn Diagram (Fig. 3)
B = most characteristic biotope (Z = epizoic, E = epiphytic, ε = epipsammic, D = diverse)

Taxon	New Caledonia NC30–NC2	French Polynesia 2010–2020 survey	Mascarenes 2005–2009 survey	B
Achnanthes cf. *brevipes* Agardh	M	M	M	D
Achnanthes pulchella Heiden	–	–	R	D
Achnanthidium glyphos Riaux-Gobin, Compère & Witkowski	R	M	M	ε
Achnanthidium pseudodelicatissimum Riaux-Gobin, Witkowski & Compère	–	–	VR	ε
**Achnanthidium sp. aff. *Achnanthes fogedii* Håkansson	–	R	M	ε
*#*Amphicocconeis antiqua* Riaux-Gobin & Coste	–	R	–	Z
*#*Amphicocconeis clypeus* Riaux-Gobin & Witkowski	–	R	–	Z
Amphicocconeis debesii (Hustedt) De Stefano	–	–	A	ε
Amphicocconeis mascarenica Riaux-Gobin & Compère	VR	A	–	ε
Amphicocconeis rodriguensis Riaux-Gobin & Al-Handal	R	–	R	ε

Taxon	New Caledonia NC30–NC2	French Polynesia 2010–2020 survey	Mascarenes 2005–2009 survey	B
*#Amphicocconeis ruatara Riaux-Gobin	–	R	–	N
Amphicocconeis sp.	VR	–	–	ε
Anorthoneis vortex Sterrenburg	–	R	R	ε
Astartiella bahusiensoides (Foged) Witkowski, Lange-Bertalot & Metzeltin	–	R	M	D
Astartiella cf. punctifera (Hustedt) Witkowski & Lange-Bertalot	–	R	R	N
*#Astartiella societatis Riaux-Gobin, Witkowski & Romero	–	M	–	N
Astartiella sp.	VR	–	–	ε
Australoneis frenguelliae (Riaux-Gobin & Guerrero) Guerrero & Riaux-Gobin	–	–	VR	D
Cocconeis alucitae Riaux-Gobin & Compère	–	VR	M	E
*Cocconeis angularipunctata Riaux-Gobin, Romero, Compère & Al-Handal	VR	M	R	ε
Cocconeis archaeana Riaux-Gobin & Compère	–	R	M	ε
Cocconeis borbonica Riaux-Gobin & Compère	R	M	A	ε
*Cocconeis borbonica 'morph'[1]	–	M	A	ε
Cocconeis cf. californica Grunow	–	VR	VR	D
Cocconeis cf. capensis (Cholnoky) Witkowski	VR	–	VR	ε
Cocconeis cf. carinata Riaux-Gobin, Ector & Witkowski	R	M	M	D
*Cocconeis convexa Giffen	M	A	A	ε
*Cocconeis convexa 'morph'[1]	R	R	–	ε
Cocconeis coralliensis Riaux-Gobin & Compère	–	M	A	ε

*Cocconeis coronatoides Riaux-Gobin & Romero 'type'	R	A	A	ε
*Cocconeis coronatoides 'morph'[1]	VR	R	–	D
*Cocconeis coronatoides 'morph'[2]	A	–	–	D
Cocconeis cupulifera Riaux-Gobin, Romero, Compère & Al-Handal	–	VR	R	ε
Cocconeis cf. delapunctata Hohn	VR	–	–	ε
***Cocconeis dapalistriata Riaux-Gobin, Romero, Compère & Al-Handal	R	VR	R	D
Cocconeis cf. dirupta Gregory	–	R	R	ε
#Cocconeis cf. dirupta var. flexella (Janisch & Rabenhorst) Grunow	–	VR	–	ε
Cocconeis diruptoides Hustedt	VR	–	–	ε
Cocconeis distans Gregory	–	A	A	D
*#Cocconeis frustrationis Riaux-Gobin, Compère & Jordan	–	R	–	ε
Cocconeis geometrica Riaux-Gobin, Romero, Compère & Al-Handal	–	R	R	ε
Cocconeis guttata Hustedt & Aleem	R	M	M	D
Cocconeis inaequalistriata Riaux-Gobin, Romero, Compère & Al-Handal	–	–	R	D
Cocconeis krammeri Lange-Bertalot & Metzeltin	–	R	R	D
#Cocconeis kurakakea Riaux-Gobin & Witkowski	–	R	–	E
Cocconeis margaritata Riaux-Gobin & Al-Handal	R	M	M	ε
Cocconeis cf. mascarenica Riaux-Gobin & Compère	M	R	A	D
Cocconeis aff. molesta Kützing	VR	A	A	D
#Cocconeis napukensis Riaux-Gobin, Compère, Coste, Straub & Taxböck	–	M	–	D
#Cocconeis ornata Gregory	–	R	VR[1]	D
#Cocconeis cf. paniformis Brun	–	R	–	ε

Taxon	New Caledonia NC30–NC2	French Polynesia 2010–2020 survey	Mascarenes 2005–2009 survey	B
Cocconeis paucistriata Riaux-Gobin, Romero, Compère & Al-Handal	VR	R	R	ε
Cocconeis peltoides Hustedt	M	A	A	D
#Cocconeis pinnata Gregory ex Greville	–	R	–	D
*Cocconeis placentula Ehrenberg Complex	A	A	R	D
Cocconeis placentula var. rodriguensis Riaux-Gobin, Romero, Compère & Al-Handal	VR	–	R	ε
Cocconeis pseudodiruptoides Foged	M	R	–	D
Cocconeis pseudograta Hustedt (+ 'morph'¹)	R	M	M	D
#Cocconeis pseudomarginata Gregory	–	R	–	D
*#Cocconeis santandrea Riaux-Gobin, Witkowski & Bemiasa	–	R	–	D
#Cocconeis sawensis Al-Handal & Riaux-Gobin	–	R	–	D
Cocconeis scutellum Ehrenberg	A	M	A	E
#Cocconeis scutellum var. posidoniae De Stefano, Marino & Mazzella	–	M	–	E
Cocconeis sigillata Riaux-Gobin & Al-Handal	R	R	A	ε
Cocconeis cf. sovereigni Hustedt	VR	–	R	D
*Cocconeis spina-christi Riaux-Gobin, Romero, Coste & Galzin	VR	R	–	ε
Cocconeis suzukii Riaux-Gobin, Compère, Coste, Straub & Taxböck	R	A	A	ε
Cocconeis cf. tsara Riaux-Gobin, Witkowski & Bemiasa	R	R	R[2]	E
*#Cocconeis tuamotuana Riaux-Gobin, Compère & Jordan	–	R	–	ε
Cocconeis sp. (cf. C. bilicis Meister)	VR	–	–	D
#Cocconeis sp. aff. rivalis Schmidt	–	R	–	D

Taxon			
#Cocconeis vaiamanuensis Riaux-Gobin, Witkowski & Ector	–	R	E
*#Cocconeis sp. 1	–	VR	D
Cocconeis sp. 2 (= ?sp. 6 in Riaux-Gobin et al. 2015c)	–	VR	D
Cocconeis sp. 3 (= sp. 5 in Riaux-Gobin et al. 2015c)	VR	VR	D
Cocconeis sp. 4 (= sp. 3 in Riaux-Gobin et al. 2015c)	–	R	D
Cocconeis sp. 5 (= sp. 4 in Riaux-Gobin et al. 2015c)	VR	VR	D
Cocconeis sp. 6 (= sp. 7 in Riaux-Gobin et al. 2015c)	–	VR	D
Madinithidium capitatum (Riaux-Gobin, Romero, Compère & Al-Handal) Witkowski, Riaux-Gobin & Desrosiers	–	R	ε
Madinithidium flexuistriatum (Riaux-Gobin, Compère & Witkowski) Witkowski, Riaux-Gobin & Desrosiers	–	R	ε
Madinithidium scalariforme (Riaux-Gobin, Compère & Witkowski) Witkowski, Riaux-Gobin & Desrosiers	–	M	ε
Navithidium pseudochamaepinnularia (Riaux-Gobin, Compère & Witkowski) Al-Handal, Romero & Wulff	R	R	ε
Planothidium delicatulum (Kützing) Round & Bukhtiyarova	–	VR	D
#Planothidium aff. frequentissimum (Lange-Bertalot) Lange-Bertalot	–	–	D
Planothidium mathurinense Riaux-Gobin & Al-Handal	–	M	ε
Planothidium rodriguense Riaux-Gobin & Compère	R	R	ε
Planothidium sp.	R	–	D
Pseudachnanthidium megapteropsis Riaux-Gobin & Witkowski	VR	R	D
Schizostauron citronella (Mann) Górecka, Riaux-Gobin, Witkowski	–	M	D
Vikingea gibbocalyx (Brun) Witkowski, Lange-Bertalot & Metzeltin	M	R	D
Vikingea florifera Riaux-Gobin, Romero, Compère & Al-Handal	VR	–	D
*#Xenococconeis opunohusiensis Riaux-Gobin, Coste & Romero	–	M	D

of enough informative-diagnostic illustration), merely assigned to the Cocconeidaceae.

4.2. Selected Achnanthales recently described from Polynesia

***Amphicocconeis antiqua* Riaux-Gobin & Coste** (Riaux-Gobin et al. 2021c). SEM, Fig. 4a–f.

Description: Valves robust, fusiform, with ogival apices. 15–24 µm long, 8–17 µm wide, n = 16. **SV**: Convex, radiate uniseriate striae, regularly spaced. SV striae 9–14 in 10 µm. Areolae oblong-elliptic, crater-like. One cupule row on each side of SV sternum. SVVC open, with short digit fimbriae. **RV**: Dense radiate striae, regularly spaced. Short and strongly bent macro-areola near the raphe, a longer one on margin. RV striae 37 ± 2 in 10 µm. Raphe filiform, straight. Central area small. Terminal raphe fissure largely hooked. RVVC extended, open, with rows of simple pores.

Type locality: Tahiti-Nui west coast (17°31.405'S, 149°31.106'W), epizoic on *Holothuria atra* Jaeger.

Holotype and etymology: *see* Riaux-Gobin et al. 2011c.

Remarks: Some differences with *A. clypeus* Riaux-Gobin & Witkowski (Riaux-Gobin et al. 2021c): SV sternum with cupules in *A. antiqua*. SVVC with short fimbriae. RV with denser striation. RVVC of both taxa with a different shape and striation. Some similarities with *A. disculoides* (Hustedt) De Stefano & Marino, but the SVVC is dissimilar, and the RVVC is open in *A. antiqua*.

***Amphicocconeis clypeus* Riaux-Gobin & Witkowski** (Riaux-Gobin et al. 2021c). SEM, Fig. 5a–d.

Description: Valves robust, oval-elliptic. 15–27 µm long. 9–14 µm wide; n = 29 (SEM). **SV**: Valves robust, striae radiate, uniseriate. SV striae 9–11 in 10 µm. Sternum narrow, straight. Areolae round to oblong, crater-like, with opening similar to a cuttlefish eye. Round warts around areola opening. On both sides of the sternum, one row of regular oblong areolae in quincunx. Marginal areolae arranged along one regular row. SVVC open, extended, with rows of round apertures. **RV**: Each stria composed of two macroareolae. Large lunar-shape space lacking ornamentation. RV striae: 27 ± 2 in 10 µm. Raphe straight. External proximal raphe endings, slightly deflected on same side. Central area small. Striae strongly radiate at mid-valve. Terminal raphe fissures hooked. RVVC extended, open, with marginal rows of small pores (27 in 10 µm), each pore row corresponding to RV stria. Often 'plow-sock' structure on head pole of RVVC.

Type locality: Sample 'Papeete 4 *Holothuria*', Tahiti Nui, Arué District, north of Tombeau du Roi (17°31.405'S, 149°31.106'W), epizoic on *Holothuria atra*.

Holotype, syntypes and etymology: *see* Riaux-Gobin et al. 2011c.

Remarks: The SV of *Amphicocconeis clypeus* shows some similarities with that of *A. catharinensis* Riaux-Gobin & Garcia (Riaux-Gobin et al. 2021c), but with some differences, such as short marginal striae in *A. clypeus*, mantle specific ornamentation in *A. catharinensis*, and very different RVVC.

Amphicocconeis ruatara **Riaux-Gobin** (Riaux-Gobin et al. 2021c). SEM, Fig. 6a–f, Fig. 7a–d.

Description: Valves elongate, slightly rostrate. Axially bent in cingular view. 14–17 µm long. 6–8 µm wide; n = 54 (SEM). **SV**: Convex, narrow concave sternum. Large, transapically elongate areolae on each part of sternum, one marginal row of short areolae and, in between, short areolae more or less in quincunx. Often low *crista marginalis* on the margin. SV striae uniseriate, slightly radiate on apices. SV striae 22 ± 3 in 10 µm. SVVC open, narrow, composed of fused fimbriae (27 in 10 µm), separated from each other by irregularly biseriate pore row, festooned edge. Very short fimbriae on SVVC head pole. **RV**: Flat, dense striae (42–60 in 10 µm) composed of narrow macro-areola, marginal striae on mid-valve. Raphe filiform. Proximal raphe endings simple, close to each other. Terminal raphe endings largely hooked. Proximal raphe endings simple, internally coaxial. Central area small. RVVC open, 30–50% valve width, with 51 rows of uniseriate pores in 10 µm.

Holotype, syntypes and etymology: see Riaux-Gobin et al. 2011c.

Locality: Ruatara rock ('Rocher de l'Homme', Raivavae), epiphyte on intertidal red turf (23°51.274'S, 147°39.595'W).

Remarks: Some similarities with *Amphicocconeis discrepans* (Schmidt) Riaux-Gobin, Witkowski, Ector & Igersheim (Riaux-Gobin et al. 2018a: figs 15–16), but with differences: The axial rows of SV areolae on each part of the sternum are larger on *A. ruatara* than on *A. discrepans*, along with, in most cases generally, a *crista margnalis*. The SVVC in *A. ruatara* has short fused fimbriae with irregularly uni- to biseriate pore rows, in contrast with a marginal row of single small pores in *A. discrepans*. The RVVC of *A. ruatara* shows dense uniseriate rows of pores (51 in 10 µm), and one row of elongate pores in *A. discrepans* (30–31 in 10 µm, Riaux-Gobin et al. 2018a).

Astartiella societatis **Riaux-Gobin, Witkowski & Romero** (Riaux-Gobin et al. 2013b). SEM, Fig. 8a–g.

Description: Cells solitary. Valves small, oblong to linear with two slight constrictions, round apices, 9–11 µm long, 2–3 µm wide; n = 34 (SEM). **SV**: Convex, striae parallel, ca. 53 in 10 µm, composed of quadrangular and dense areolae with granular hymenes. One apical row of elliptical areolae. Virgae equidistant. Axial area narrow, straight. Several large cingular bands without perforations. SVVC open. Vestigial raphe. **RV**: Flat to slightly concave. Striae parallel, ca. 38 in 10 µm. Striae equidistant. Round areolae closed by finely granulate hymenes. One small stigma (or fistula), on secondary side of central area, axially elongated and internally closed by domed hymen. Central area discrete, asymmetrical and wider towards stigma. Raphe straight, proximal raphe endings simple, close to each other, slightly bent toward stigma, internally coaxial. Terminal raphe fissures doubly hooked towards same side. RV areolae internally closed by perforated hymenes.

Holotype, isotypes and etymology: *see* Riaux-Gobin et al. 2013b.

Type locality: Sample 'Papeete 4 Holothuria', Tahiti Nui, Arué District, north of Tombeau du Roi (17°31.405'S, 149°31.106'W), epizoic on *Holothuria atra* Jaeger.

Remarks: Some similarities with *Astartiella bremeyeri* (Lange-Bertalot) Witkowski & Lange-Bertalot (Witkowski et al. 2000). In LM, resembles *Achnanthidium crassum* (Hustedt) Potapova & Ponader, which is broader, has a RV with a symmetrically expanded central area and lacks a stigma (Potapova & Ponader 2004). *Astartiella societatis* is relatively common on small coral sand grains found in a holothurian scraping, but an obligate epizoic character remains to be established. Thus far, only observed as epizoic in Tahiti and never in the nearby marine or freshwater sediments.

Cocconeis frustrationis **Riaux-Gobin, Compère & Jordan** (Riaux-Gobin et al. 2015a). SEM, Fig. 9a–f, Fig. 10a–d.

Description: Small, elliptic to sub-round valves. **SV**: Robust, convex. Striae radiate (n = 24; 9–11 µm long, 2–3 µm wide; SV 14 ± 1.0 in 10 µm), uniseriate close to sternum, biseriate but not alternating near margin, last row of areolae located in a sort of groove. Striae composed of reduced number of sub-quadrangular large areolae, axially slightly elongated, forming longitudinal rows. Areolae occluded externally by finely perforated hymenes, internally ornamented with small crystal-like granules. Regular interstriae. Sternum straight, relatively narrow and linear. SVVC robust, closed, with regular and short fimbriae lying on each interstria. SVVC abvalvar side with median groove. SVVC advalvar side with sunken areas corresponding to end of short SV 'chambers' laterally delineated by reinforced interstriae. **RV**: Thin, flat. Striae uniseriate (SV 23 ± 2.9 in 10 µm), radiate. Regular interstriae. Presence of hyaline rim well apart from margin. Round areolae, externally occluded by hymenes apparently lacking linear perforations. Raphe filiform, straight. Terminal raphe endings simple, far from valve margin and ending before hyaline rim. Robust and closed RVVC with long and thin spine-like fimbriae (> 1/4 of valve width) irregularly positioned, with lateral basal expansions like a quill pen. On abvalvar side, each RVVC fimbria has elongate short crest without transverse slots. One fimbria at each RVVC apex. Cingulum shallow with no additional bands.

Type locality: Sample '15NPK7–2', Napuka Atoll (14°10.702'S, 141°15.949'W).

Holotype, isotypes and etymology: *see* Riaux-Gobin et al. 2015a.

Remarks: Contrarily to *C. scutellum* Ehrenberg, the RV striae are uniseriate up to the margin. Instead of a papilla with transverse furrows on each fimbria of the RVVC, there is an oblong short crest. Also differs from *C. coronatoides* Riaux-Gobin & Romero, *Cocconeis frustrationis* and *C. spina-christi* Riaux-

Gobin, Romero, Coste & Galzin (Riaux-Gobin et al. 2013b) *see* Riaux-Gobin et al. 2015a.

Cocconeis kurakakea **Riaux-Gobin & Witkowski** (Riaux-Gobin et al. 2018d). SEM, Fig. 11a–e.

Description: Small valves, linear-elliptic (7–16 µm long), narrow (3–6 µm wide); n = 54 (SEM). SV sternum and raphe sigmoid. SV and RV striation almost indiscernible with LM. **SV**: Striae (47–54; SV 49 ± 1.3) regularly spaced, only slightly radiate on apices. SV areolae small, slightly elongate and decussate. One marginal row of elongate areolae. No hyaline marginal area. SV sternum narrow and sigmoid. SV areolae with round to oblong internal foramina. Marginal SV areolae covered with lateral foramina. SV marginal short striae present on apices. SVVC as open and imperforated large band, without fimbriae. Cingulum composed of 4 narrow open copulae, up to 6. **RV**: Valve weakly silicified, flat to slightly concave. Striae (38–42; RV 40 ± 1.2) regularly spaced, parallel to radiate close to apices, composed of small round areolae. One row of marginal slightly elongate areolae. Raphe sigmoid, proximal raphe endings straight, simple and close together. Short hemi-fascia, sometimes lacking. Axial area narrow. RV areolae externally closed by hymenes with radiate slits. RVVC open, without fimbriae. Terminal raphe endings bent in opposite directions. No areolae on apices. With SEM, species easily distinguishable by SV areola structure and pattern, RV hemi-fascia and high cingulum.

Holotype and etymology: *see* Riaux-Gobin et al. 2018d.

Type locality: Nukutavake reef (Tuamotu Archipelago, 19°16.833'S; 138°47.117'W). Sample 'NTV A1'. Oct. 2016.

Habitat: Undetermined filamentous green macroalgae on coral reef slope. Until now only found on Nukutavake filled atoll.

Remarks: Some similarities with *Cocconeis diruptoides* Hustedt (1933). See discussion in Riaux-Gobin et al. (2018d). Also, some similarities with *Cocconeis pseudonotata* De Stefano & Marino (De Stefano & Marino 2001), but the SV areolae of *C. pseudonotata* are 'alveolae' composed of loculi and its RV fascia reaches the margins. *Cocconeis pseudodiruptoides* Foged (Foged 1975; illustrations from the type slide Pla 62a (housed in Copenhagen (C), Herbarium, Botanical Museum, Copenhagen, Denmark) may also be compared to the new taxon, but it has an extended fascia on both valves, and the SV areolae are complex loculi. Two taxa, recently described as *C. caulerpacola* Witkowski, Car & Dobosz (Car et al. 2012) and *C. borbonica* Riaux-Gobin & Compère (2008), also have some similarities with *C. kurakakea*.

Cocconeis santandrea Riaux-Gobin, Witkowski & Bemiasa (Riaux-Gobin et al. 2021b). SEM, Fig. 12a–e.

Description: Valves robust, elliptic. 16–24 µm long, 12–16 µm wide; n = 25 (SEM). **SV**: Convex. Striae radiate, regularly spaced, uniseriate on valve face, becoming quadriseriate with alternate areolae in last 1/3 of valve and mantle. SV striae 9–13 in 10 µm. SV with regular axial rows of slightly squared areolae, externally concave, opening internally by round foramen. Loculate areolae concave, cross-shaped, delineated by four small pegs in axial and transverse positions that extend towards center of areola, but never merging in center of areola. Each peg frequently externally ornamented by one or two round warts. Hymenes perforated by small pores, and short slits on periphery. SVVC with digitate fimbriae regularly aligned with SV striae. **RV**: Concave, with a marginal hyaline rim positioned far from valve margin, striae uniseriate up to margin and biseriate with alternate areolae afterwards, with specific 'ears of wheat' pattern. One row of round and larger areolae close to hyaline area. Elevated internal rim with bumps. RV striae 15–18 in 10 µm, composed of regularly spaced, small and round areolae. Areola hymenes with radiate slits. Raphe straight. Central area small. External proximal raphe endings closely spaced. RVVC with fimbriae irregularly spaced, and joined at their edge, creating large rectangular fenestrae. RVVC papillae oblong, with up to 5 furrows.

Holotype and etymology: *see* Riaux-Gobin et al. 2021b.

Type locality: Intertidal turf, Rapa (Austral, South Pacific), sample 'Rapa–1'. Also found in Nuku-Hiva (Marquesas) on intertidal turf.

Habitat: Epipsammic on coral sand grains and epiphytic on short turf from coral reef environments in the South Pacific.

Remarks: *Cocconeis santandrea* can be distinguished from *C. nosybetiana* Riaux-Gobin, Witkowski, Bemiasa & Bemanaja (Riaux-Gobin et al. 2019c) by the structure of the loculate SV areolae. The X-shaped areolae of *C. santandrea* are split into 4 sectors, while those of *C. nosybetiana* are transversely split into two parts. The valve outline of *C. santandrea* is oblong-elliptic versus round-elliptic in *C. nosybetiana*.

Cocconeis spina-christi **Riaux-Gobin, Romero, Coste & Galzin** (Riaux-Gobin et al. 2013a). SEM, Fig. 13a–f, Fig. 14a–f.

Description: Valve small, elliptical. **SV**: 12–22 µm long, 9–14 µm width. Externally, valve surface convex and strongly silicified. Sternum narrow, linear. Striae radiate, uniseriate close to sternum and bi- to triseriate near margin, with lower stria density than in RV. **SV**: 12–18 in 10 µm; n = 38 (SEM). Round areolae (12–16 in 10 µm), regularly axially arranged, also present at apices. One apical row of smaller areolae. Areolae externally occluded by hymenes with linear perforations, radiating from depressed central part. SVVC robust and closed, digitate fimbriae. Abvalvar side of SVVC shows ovoid to ellipsoid cupules, more or less regularly arranged all over margin. **RV**: Weakly silicified. Striae dense (20–30 in 10 µm), regularly spaced and radiate, uniseriate up to marginal hyaline rim. Round and small areolae (26–35 in 10 µm). RV striae biseriate with alternating areolae between rim and valve margin. Central area reduced. Axial area narrow-linear. Raphe filiform, straight, proximal raphe endings approximate and not bent, distal raphe endings simple. Raised, narrow hyaline marginal rim. Cingulum with no supplementary copulae. RVVC thick, well developed and closed, with marginal subquadrangular to ovoid large fenestrae formed by fusion of two narrow fimbriae. On abvalvar side of each fimbria, is a rounded and elevated papilla, ornamented with three to six furrows. Inner edge of RVVC with spine-like fimbriae of different size and location. At each apex of RVVC, usually, a larger spine, sunken on its advalvar side.

Holotype, isotypes and etymology: *see* Riaux-Gobin et al. 2013a.

Type locality: Sample 'COOK1', intertidal beach sand (Cook Bay, Moorea Island, Society, 17°29.600'S, 149°49.300'W).

Cocconeis tuamotuana **Riaux-Gobin, Compère & Jordan** (Riaux-Gobin et al. 2015a). SEM, Fig. 15a–f.

Description: Valve elliptical to broadly elliptical in bigger specimens, round apices.

SV: Convex, bi-layered, with two longitudinal rows of alveoli (up to three in biggest specimens), sternum sigmoid and concave, constricted in its mid-part. Two more or less elongated crescent-shaped hyaline areas, concave, on each side of sternum, delineating two sets of alveoli. Striae uniseriate, 29–32 str. in 10 µm. Alveoli regularly radiate, externally open via oblong narrow lumina. Apical alveoli short and deflected in opposite directions. Each alveolus opens internally by small round foramen, along regular apical lines. SVVC very thick, open, lacking fimbriae. Cingulum composed of two open valvocopulae and additional connecting copula with ligula. **RV**: Slightly concave externally, with two embossed lateral areas lacking some areolae. Central area enlarged and round, no marginal rim. Striae uniseriate (21–25 str. in 10 µm). Areolae small, closed by hymenes with very short marginal slits. Axial area narrow. Raphe strongly sigmoid. Approximate, straight central raphe endings, internally deflected in opposite directions. Terminal raphe endings strongly curved in opposite directions. RVVC large, with abvalvar irregular edge. Regular embossed designs (corresponding to each RV areola) on advalvar side of RVVC closely fitting onto the RV areola structure. At apex, RVVC with bent hole corresponding to location of helictoglossa.

Holotype, isotype and etymology: *see* Riaux-Gobin et al. 2015a.

Type locality: Sample '15NPK7–2', Napuka atoll (14°10.702'S, 141°15.949'W).

Remarks: The smaller specimens of *C. tuamotuana* closely resemble *C. paniformis* Brun (Schmidt 1874–1959, Montgomery 1978), later described as *C. caribensis* Romero & Navarro (1999). The internal SV illustration *in* Romero & Navarro (1999) is very similar to ours, but the SVVC lacks fimbriae. Nevertheless, neither the RV nor the SV of the Napuka taxon has a fascia. The biggest specimens from Napuka bear some resemblance to *C. heteroidea* Hantzsch which seems to be a polymorphic species (Schmidt 1874–1959, Suzuki et al. 2001, De Stefano & Romero 2005). Nevertheless, a significant difference from the latter is the presence of a simple row of foramina on each side of the SV in our specimens, up to three in the biggest specimens, whereas there are multiple rows in *C. heteroidea* (Suzuki et al. 2001). In addition, the SV embossed axial design on each part of the SV sternum is absent in the Napuka taxon. The two RV embossed lateral areas lacking some areolae, the slightly crenulated RVVC, and the small central area seem also unique to *C. tuamotuana*. Although the overall design of the Napuka taxon fits with *C. heteroidea*, the above substantial differences allow the recognition of *C. tuamotuana*.

Cocconeis vaiamanuensis Riaux-Gobin, Witkowski & Ector (Riaux-Gobin et al. 2021d). SEM, Fig. 16a–f.

Description: Valve oblong-elliptic to linear (7–15 µm long, 3–5 µm wide; SEM, n = 60), round apices. **SV**: Slightly concave, narrow sternum. Small round areolae (4–5 in 1 µm). SV striae uniseriate (16–22 in 10 µm), parallel in mid-valve to radiate and slightly denser on apices. Striae composed of tiny areolae, more or less in zig-zag, internally closed by strongly convex hymenes without obvious slits or punctuations. Mantle narrow with one row of small pores providing access to a process. These small processes are internally closed by a hemispheric plug or hymen, slightly different and smaller in size than that of SV areolae. One process faces each stria. Strong virgae, externally embossed on their most marginal part. High cingulum composed of several open and large cingulae devoid of ornamentation. SVVC with no fimbriae. **RV**: Strongly convex. Striae radiate, slightly denser on apices (20–24 in 10 µm), small areolae, biseriate near axial area, uniseriate on a short median section, and up to quadriseriate near margin. Striae distinctly biseriate on rare specimens, with alternate areolae. Marginal row of dense, apically elongate areolae (60–72 in 10 µm), separated from the rest of valve by large hyaline apical area, corresponding to internal flat unraised rim. Central area absent or very small (short hemi-fascia on some specimens). Raphe filiform, straight. Proximal raphe endings externally close to each other, slightly bent on same side, internally bent in opposite directions. Terminal raphe endings simple, close to margin, apically surrounded by anchor-like silica fold with no connection to valve interior. RVVC with slightly undulated edge. Central area almost lacking. Helictoglossa straight and unraised.

Holotype and etymology: *see* Riaux-Gobin et al. 2021d.

Type locality: Ruatara rock ('Rocher de l'Homme'), Raivavae (Austral), sample 'RAI 20' (intertidal red macroalgal turf). 23°51.274'S, 147°39.595'W.

Ecology: Relatively rare, living on rocky shore covered by turf.

Distribution: Until now only found on Raivavae.

Remarks: Relatively rare taxon. Some dissimilarity with *Cocconeis nugalas* Hohn & Hellerman (Hohn & Hellerman 1966) [synonym *C. hauniensis* Witkowski (Witkowski 1993; Desianti et al. 2015; Riaux-Gobin & Romero 2003]. The two taxa are significantly dissimilar (Riaux-Gobin et al. 2021d). Based on LM, similarities also exist with *Cocconeis finmarchica* Grunow *in* Cleve & Grunow (Cleve & Grunow 1880). The striation on both valves of *C. finmarchica* looks quite similar to that in *C. vaiamanuensis*, but differs from the latter by having a narrow and long fascia on the RV, and a frustule shape that is ellipsoid. No recent bibliography includes SEM photographs of *C. finmarchica*.

***Xenococconeis opunohusiensis* Riaux-Gobin, Coste & Romero** (Riaux-Gobin et al. 2014b). SEM, Fig. 17a–d, Fig. 18a–e.

Description: Valves elliptical, round apices, 10–25 µm long, 7–14 µm wide, n = 77 (SEM). **SV**: convex, with elliptic sternum (1/3 of valve width) axially depressed. 11–16 SV striae in 10 µm, short, radiate and equidistant, uniseriate near sternum, biseriate to triseriate near margin. 30 small areolae in 10 µm, round to irregularly oblong, occluded by depressed hymen with radiate slits, internally domed. Irregularities in SV striation often noted, corresponding to position of underlying ribs of valvocopula. SVVC extended, with transapical ribs: 5–6 in 10 µm, slightly denser on apices. On advalvar side, SVVC with extended pars interior composed of large and fused digit fimbriae. Near margin, the fimbria edges fit SV areolae. Each peg-like fimbria edge lying on SV virga. SVVC with a central elliptical large foramen. SVVC on abvalvar side, with robust marginal and short concave ribs ornamented by lateral round and small bosses, managing pseudoloculi. These ribs act as female part of a snap fastener connected to raised oblong papillae located on RVVC. SVVC often found detached. The SVVC ribs locate on two SVVC fimbriae, while the roof of the pseudoloculi lies on two SV striae and one SVVC fimbria. **RV**: Plane to slightly convex, loosely silicified. 22–28 striae in 10 µm. 45–55 areolae in 10 µm. Striae radiate, regularly spaced, uniseriate with irregularities to biseriate on margin, with in quincunx oblong areolae. Several apical short biseriate striae separated from terminal raphe endings by small hyaline area. Areolae round to subquadrangular, occluded by hymen close to valve face, or slightly depressed, with radiate slits arranged in sectors. Axial area straight and narrow. Central area sub-round and reduced. Raphe straight and filiform. Central raphe endings externally very close. Distal raphe endings simple, well apart from margin. RVVC extended, with marginal elliptic fenestrae (av. 5 in 10 µm), denser on apices. with an extended pars interior with elliptic marginal fenestrae. Multiform foramen in center of RVVC: from restricted and round, to half-valve wide and subquadrangular to highly irregularly shaped. In mid-valve, floor of RVVC is plain or opened by one round foramen between each rib. On abvalvar face, RVVC bears raised regular marginal short and sharp ribs, extending from RV margin to mid-valve, ornamented by finely ribbed and vermiform papilla. Papilla acts as male structure, exactly fitting concave structure of opposite SVVC rib. Median and also marginal lines of embossed dots, reinforcing anchorage of RVVC to RV. RVVC structure regularly organized. Cingulum relatively high. SVVC with complex ultrastructure (see original description and illustrations).

Type locality: Opunohu Bay, Moorea (17°31.047'S, 149°51.021'W).

Holotype, isotype and etymology: *see* Riaux-Gobin et al. 2014b.

4.3. New details on two pantropical taxa

Cocconeis angularipunctata **Riaux-Gobin, Romero, Compère & Al Handal** (Riaux-Gobin et al. 2011c). SEM, Fig. 19a–d, Fig. 20a–d.

The specimens of *Cocconeis angularipunctata* in our Pacific material added new data to the original RV description from Mascarenes (Riaux-Gobin et al. 2011c, pl. 29/1–6). Clumps of this taxon were observed on a turf containing coral detritic particles. Complete frustules, broken specimens with both valves, and RV still in place (with their internal side) were observed, all specimens side by side (sample 'Papeete 3b', Taiharuru, 24 10 2010, 17°31.431'S, 149°31.233'W). Relatively rare in the Pacific (14 specimens observed, SEM, 5–11 µm long, 3–6 µm wide. SV: 28–33 str. in 10 µm. RV: 33–37 str. in 10 µm). Figure 19a–d illustrates the SV with marginal long areolae bordered by slits, and RV with a marginal hyaline rim (Fig. 20a–d). RVVC with undulated edge and SVVC with no fimbriae.

Remarks: The Pacific specimens of *C. angularipunctata* only differ from those from Mascarenes by a slightly denser SV striation. The RV has some similarity with that of *C. mascarenica* Riaux-Gobin & Compère, with same stria density and a hyaline marginal rim, whereas the SV areolae (Fig. 18a,c,d) are different from those in *C. mascarenica*, that are narrow and slightly denser.

(**Holotype, isotypes, etymology, type locality**: see Riaux-Gobin et al. 2011c).

Cocconeis geometrica **Riaux-Gobin, Romero, Compère & Al-Handal** (Riaux-Gobin et al. 2011b). SEM, Fig. 21a–f.

Originally described from the Mascarenes (Riaux-Gobin et al. 2011b, p. 26–27, p. 144, pl. 45/1–6). Some Pacific specimens of *Cocconeis geometrica* permitted a better illustration of its RV external view (Fig. 21e,f). Rare (10 specimens observed in our Pacific samples, SEM: 8–11 µm long, 3–5 µm wide. SV: 28–30 str. in 10 µm. RV: 28–36 str. in 10 µm). RV striae equidistant and radiate on apices. Central area reduced. No marginal hyaline rim, areolae with short radiate slits. Terminal raphe endings far from the margin (Fig. 21f). RVVC open (Fig. 21d), with short trapezoidal fimbriae of irregular length. Helictoglossa slighty raised and gutter-shaped (Fig. 21d).

Remarks: The biometrics of the Pacific specimens of *C. geometrica* only differ from those of the Mascarenes individuals by a slightly denser striation on both valves. The RVVC, with short fimbriae of irregular length, as well as the geometrical shape, position and arrangement of the SV areolae, are unique.

(**Holotype, isotypes, etymology, type locality**: see Riaux-Gobin et al. 2011b).

4.4. Rare and unnamed tropical taxa

Unnamed small and rare Achnanthales taxa were recorded during the 2010–2020 Pacific survey (Table 1, *Cocconeis* sp. 1–6), several of which were found throughout the whole Indo-Pacific Basin, while others seemingly restricted to the Pacific. Among the latter, *Cocconeis* sp. 7 (Fig. 22e–f), not listed in Table 2, was only observed as very rare from Raivavae (RV not observed). 23 equidistant SV striae in 10 µm, composed of round small and equidistant areolae (30 in 10 µm) with irregular alignment (Fig. 22e), a marginal row of simple processes (Fig. 22f). This taxon pertains to the group of taxa with simple processes (Riaux-Gobin et al. 2015). Some similarities with *C. peltoides* Hustedt, but with denser SV striae (23 versus av. 13 in *C. peltoides*), lacking the mid-SV axial structure splitting the valve in two parts, and with no *crista marginalis*. We present in the same plate (Fig. 22a–d) another taxon pertaining to the same group, as *Cocconeis* sp. 6 (cited as (?)*Cocconeis* sp. 7 *in* Riaux-Gobin et al. 2015, figs 9, 35–38, Sub-group 1C striae composed of very few areolae). The latter taxon was epizoic on *Holothuria atra* (sample 'Papeete 4 holoth'). Its presence in Mascarenes has to be confirmed. The RV of *Cocconeis* sp. 6 is unknown 'so, its ascription to *Cocconeis* remains doubtful', but it has some similarities with *Cocconeis pseudograta* Hustedt, with SV 'raised virgae in place of nodules, and no crista marginalis'. The SV areolae, in internal view are bordered by small short radiating indentations, quite unusual. The marginal simple processes are internally opening on the top of a dome, and the central hole is bordered by a regular row of small siliceous beads (Fig. 22d).

Among the rare taxa apparently unique from Polynesia, we can also cite *Cocconeis* sp.1, probably pertaining to the *C. scutellum* Complex (Fig. 23a,d), detailed in '*Cocconeis scutellum* Complex').

4.5. Morphotypes or formae

Several taxa observed during the present survey had a high morphological variability if compared to the original description. The name 'morph' herein refers to slight variations, e.g. concerning the frustule shape: some specimens of *Cocconeis convexa* Giffen (1967) have an oblong valve shape (Fig. 24a–f), while others are round to sub-discoid, fitting the original description (Fig. 25a–d). The latter oblong 'morph' differs from *C. pediculus* Ehrenberg by lacking arborescent fimbriae on the SVVC, and by simple SV areolae, while *C. pediculus* has complex ones (cf. SEM images in Spaulding et al. from https://diatoms.org). Specimens with long SV areolae and a restricted number of axial rows of areolae (Fig. 24d) may refer to another taxon. Oblong 'morphs' of *C. pseudograta* Hustedt (1939) were also observed during this survey (not illustrated). The variations can also affect the overall dimensions of taxa, i.e., some specimens of *C. guttata* Hustedt & Aleem (1951) were much bigger than currently observed, with a larger SVVC, and may be considered as a 'morph' (not illustrated).

Variability may also concern the areola arrangement. The Pacific specimens of *Cocconeis borbonica* Riaux-Gobin & Compère have, for the main part, one row of obviously longer SV areolae on each side of the SV sternum (Fig. 26), while in the original diagnosis (Riaux-Gobin & Compère 2008) it was stated 'oblong areolae regularly arranged into four or five longitudinal rows decreasing in width near the margin'. The Pacific *C. borbonica* 'morph'1 (Table 2) can be compared to *C. caulerpacola* Witkowski, Car & Dobosz (Car et al. 2012), a taxon created later on, with no significant morphological differences from *C.*

borbonica. In the original description of *C. borbonica* (ref. cit., figs 29–30) the illustrated specimens have a slightly sigmoid raphe, even if not explicitly reported in the original diagnosis, we commented that the 'terminal raphe endings' are 'slightly deflected in opposite side' (Riaux-Gobin & Compère 2008).

A great variability was also observed in *Cocconeis coronatoides*, as pointed out when commenting on the Melanesian assemblages from New Caledonia (Riaux-Gobin et al. in press). In Polynesia, *C. coronatoides* is observed with several formae or 'morphs', with rare and almost discoid large specimens ('morph'1, Table 2, Fig. 27a,b,d), and some specimens with an oblong to linear shape (Fig. 27c,e,f) close to the original description (Riaux-Gobin et al. 2010). A 'morph' of *C. coronatoides*, recently reported off New Caledonia (ref. cit.), can be slightly different from the two forms encountered in Polynesia, with a smaller length, no SV *crista marginalis*, and almost no silica beads spread over the SV face ('morph'2, Table 2). The presence of different 'morphs' in *C. coronatoides* among assemblages, may be related to environmental conditions (epiphytic or epipsammic ethology), water quality (calcareous or silicic, with a more or less high conductivity) in possible relation-connexion with the geologic past of each island. A genetic analysis of these 'morphs' would prove if some are unique and 'potential endemic' taxa (as 'cryptic taxa') or if they are all conspecific, only witnessing unique environmental conditions. Note that *C. coronatoides* has been misidentified in the past, i.e. as *C. placentula* Ehrenberg in the British West Indies, Grand Cayman (South of Cuba Island, Caribbean Sea, *in* Corlett & Jones 2007, fig. 4B), and illustrated as *Cocconeis* sp. 4 *in* Montgomery (1978, pl. 60EFG).

4.6. Ill-defined taxa

4.6.1. *Cocconeis placentula* Complex

During the past decades several groups of *Cocconeis* gave rise to poor definitions. This is particularly the case with the '*C. placentula* Complex'. Among this group, *Cocconeis placentula* Ehrenberg 1838, *C. lineata* Ehrenberg 1849, as well as *C. euglypta* Ehrenberg 1854, and afterwards some new combinations such as *C. lineata* var. *euglypta* (Ehrenberg) Grunow 1880 (Grunow *in* Van Heurck 1880) and *C. placentula* f. *euglypta* (Ehrenberg) Hustedt 1957, are all taxa generating a lot of debate. Potapova & Spaulding (2013) commented (https://diatoms.org/species/cocconeis_placentula) 'yet further studies are necessary to determine whether and how individual species within *C. placentula* sensu lato may be distinguished using morphological characters'. Marina Potapova (site cit.) gave illustrations of specimens with short and dense SV areolae with no particular arrangement, along with specimens with SV longer areolae disposed along regular axial rows.

The original drawing selected by Romero & Jahn (2013, fig. 1) as lectotype for *Cocconeis lineata* (original illustration for *C. lineata* from Ehrenberg collection, drawing entitled *C. lineata* by Ehrenberg in his drawing sheet 2177, from 1849), and the drawings selected by Tuji (2009, fig. 3e,f), also from Ehrenberg original drawings, published later on (Ehrenberg 1854, pl. 6, fig. 40 and pl. 9-I, fig. 47), apparently show entire frustules, clearly with regular axial and regular rows of ornamentations. These transapical rows, or 'lines' (*linea* in Latin), are less that five on each hemivalve. In contrast, the lectotype chosen by Romero & Jahn (2013, fig. 9) for *Cocconeis euglypta* is an LM illustration of an SV from the original Ehrenberg mica material, with no real axial rows, but in zig-zag broken lines. The drawings selected for *C. placentula* by Tuji (2009, fig. 3g,h, from Ehrenberg 1843, and fig. 3 I, from Ehrenberg 1854), do not show any axial rows of ornamentation, and the lectotype illustrated by Romero & Jahn (2013, fig. 1), for *C. placentula* var. *placentula*, is not indicative at all.

None of the above cited original materials is available for SEM, so the real or putative ultrastructure of a particular taxon is uncertain. Nevertheless, the original latin denomination '*lineata*' would imply an obvious presence of 'regular lines' perfectly visible even through light microscopy used at the time of its discovery (1834), so one would be inclined to name *Cocconeis lineata* a species with obvious SV axial rows of areolae, such as illustrated in Fig. 28. Nevertheless, there is a continuum among our Pacific specimens, with some of them showing zig-zag axial rows of SV areolae, even if still quite long areolae (not illustrated), in contrast with some others, where they are often larger in size and less slender, with short SV areolae with no axial pattern (Fig. 29). Thus, and following Potapova & Spaulding (2013, https://diatoms.org/species/cocconeis_placentula), we herein group all these taxa under *C. placentula* sensu lato, bearing in mind that it may be grouping different taxa.

The clones designated as epitype by Jahn et al. (2009) and Romero & Jahn (2013), for *C. placentula* var. *placentula* (monoclonal strain D36_012, Berlin, Germany), *C. lineata* (strain 17_011, Faroe Islands) and *C. euglypta* (strain WiCoc02, Salakchopko River, Florida, USA), have no connection to the original type materials, nor to the type localities, and as such may be questionable. Monnier et al. 2007, illustrating and commenting on old drawings and diagnoses, proposed a morphological SV differentiation between *C. euglypta* and *C. lineata*, and concluded that the morphological difference may be related to the environment. Tuji (2009), working on the original material in Ehrenberg collection, particularly EC 2225, the origin of Ehrenberg's drawings (1854, reproduced *in* Tuji 2009), concluded that 'Although the range in valve size is very large, the morphological variation'…'seems to be the variation of one taxon. Ehrenberg may have distinguished the raphid valve and araphid valve as different taxa'. Tuji (2009), therefore, did not envisage two taxa.

Cocconeis placentula, often cited in freshwater environments, is a polymorph and ubiquitous taxon also present in marine environments. All specimens illustrated from our Polynesia samples (Figs 28–29) show a well identifiable RV hyaline rim. Nevertheless, it is important to focus not only on the latter rim, but also on the valvocopulae. The SVVC and the RVVC of these taxa are rarely illustrated or commented on, there may be a strong reason for giving them a taxonomic position. In the description of the epitypes by Jahn et al. (2009) and Romero & Jahn (2013), the valvocopulae are poorly illustrated, with no comments.

Illustrating the importance of the valvocopula ultrastructure, a marine taxon from the Mascarenes was described as *C. placentula* var. *rodriguensis* Riaux-Gobin, Romero, Compère & Al-Handal (Riaux-Gobin et al. 2011c). This taxon has an unclear (not well-marked) RV marginal rim, but possesses all the other morphological characters of *C. placentula*. A main distinctive feature of the latter species was its RVVC with long and irregular fimbriae, each with several round papillae. The latter taxon was present in New Caledonia (Melanesia, Riaux-Gobin et al. 2022b & Table 3).

Regarding freshwater taxa showing similarities to the *Cocconeis placentula* Complex, a recently created temperate taxon, *C. cascadensis* Stancheva, is described as 'possessing a unique combination of morphological features, representing an intermediary between the more complex valve ultrastructure of marine *Cocconeis* taxa and the freshwater members of the *C. placentula* species Complex, with simple slit-like areolae' (Stancheva 2019).

Furthermore, Mora et al. (2021) recently documented the morphological differences linked to the reproductive cycle of *C. czarneckii* Stancheva, Mora & Jahn. The latter taxon 'shares' morphological 'similarities with *C. sijunghoensis* Jahn & Suh (*in* Jahn et al. 2020), *C. amerieuglypta* Costa, Wetzel & Ector (2020), *C. klamathensis* Sovereign (1958) and *C. grovei* Schmidt' (ref. cit.), and pertains to the *C. placentula* Complex. The RV of *C. czarneckii* shows a great variability, with more or less absent hyaline external rim (ref. cit., fig. 53, 54), or a perfectly identifiable and well-marked one (ref. cit., fig. 56, 57). Immediate post-initial cells seem to have a unique morphology. Note that individuals such as those illustrated in figs 47, 48, and 56 (ref. cit.), with the RV rim very far from the valve margin, have high similarities with *C. klamathensis* (see Spaulding et al. Diatoms of North America. Retrieved January 21, 2022, https://diatoms.org/species/cocconeis_klamathensis). A cladistic analysis built from their table 3 (ref. cit.) would tentatively illustrate the relationships between all these taxa, and indicate their degree of similarity inter se.

4.6.2. Cocconeis scutellum **Complex**

The '*Cocconeis scutellum* Ehrenberg Complex' is highly diversified (see Mizuno 1987, De Stefano et al. 2008, Riaux-Gobin et al. 2018b), with numerous varieties that would merit a re-examination after checking the original type material when available, and using modern morphological tools such as the SEM. The actual tendency is to create new species in place of varieties (e.g. new species from the *C. placentula* Complex *in* Jahn et al. 2020, while their morphology can only be disentangled with SEM). Within the *C. scutellum* Complex, in assemblages from all over the world, the bibliography is also complicated, with taxa often listed without any illustrations, and with no reference to original materials.

Cocconeis scutellum var. *parva* (Grunow) Cleve is such an example. The latter taxon is cited by several authors (see review *in* Riaux-Gobin et al. 2018b, table 3), and also, with no illustration by Hendey (1971, Galápagos), Foged (1984, Cuba), and Frankovich & Wachnicka (2015, Florida Bay). After examination of a slide from Grunow's slide collection N° 1035 (leg. by Lindig no 96) (no raw material left for SEM), it seems that the taxon was 'inaccurately described in several of the previously cited works (ref. cit.), particularly for those describing SV loculi (Okuno 1957, Suzuki & Kobayashi 2002, Suzuki et al. 2005)', or rotae (Zupo et al. 2014). The taxa with SV loculi may refer to entities pertaining to the Loculatae (see 'Loculatae section'). Following Riaux-Gobin et al. (2018b), *C. scutellum* var. *parva* is part of a *continuum* within the nominate variety.

Cocconeis scutellum var. *ornata* Grunow (1868) also led to many misidentifications in the past (i.e. Riaux-Gobin et al. 2019, table 1). Lobban et al. (2012, pl. 42, figs 4–5) cited this taxon from Guam (West Micronesia) and illustrated it with 'two oblong-elliptical SV with complex loculi'. The latter Guam taxon may be close to *Cocconeis* cf. *tsara* Riaux-Gobin, Frankovich, Witkowski, Saenz-Agudelo, Esteve, Ector & Bemiasa encountered on the green turtle 'Océane' (Riaux-Gobin et al. 2021b, figs 30–33, Haapiti, Moorea Island). *Cocconeis tsara* pertains to the Loculatae section (see next chapter). *Cocconeis pseudornata* Riaux-Gobin, Igersheim, Ector & Witkowski (Riaux-Gobin et al. 2019b), was described after SEM examination of the type material of *C. scutellum* var. *ornata* (material from capsule N° 131–6, annotated by Grunow as containing the species). To our knowledge the latter taxon was not observed in the Pacific.

Of the several taxa observed during the present survey, and unnamed due to insufficient documentation, is a small taxon from Moorea Island, from the sample 'Cook1beach' (two SV illustrated, no RV observed), with a unique discontinuous SV *crista marginalis* and with silica beads more or less evenly distributed over the valve (Fig. 22). The latter taxon may be related to the *C. scutellum* Complex, with marginally biseriate SV striae and a unique arrangement (Fig. 22c, arrows). This taxon has simple SV areolae with radial slits (no loculi). Both valvocopulae and the RV need to be found before formal classification. Another taxon, with all morphological characters of *C. scutellum*, but with RVVC fimbriae with cupuliform round papillae (in place of elongate papillae with transversal furrows as usally reported), has been identified as not rare from Tahiti-Nui (fig. 49 *in* Riaux-Gobin et al. 2018b), but may be a simple environmental 'morph' ('varieties with phenotypic expression' ref. cit.).

4.6.3. Loculatae section

Several taxa pertaining to the *C. scutellum* Complex, or previously defined as close to this group, have complex SV areolae, named 'loculi', of which *C. scutellum* var. *posidoniae* De Stefano, Marino & Mazella (De Stefano et al. 2000), and more recently described species, such as *C. nosybetiana* Riaux-Gobin, Witkowski, Bemiasa & Bemanaja (Riaux-Gobin et al. 2019b), *C. tsara* (Riaux-Gobin et al. 2021b) and *C. santandrea* (described from Rapa Island, ref. cit.) are members. A preliminary cladistic analysis for taxa pertaining to the Loculatae section is presented *in* Riaux-Gobin et al. (2021b). *Cocconeis speciosa* Gregory (1855) (cited from Revillagigedo, close to the Mexico coast, by Siquieros-Beltrones et al. 2021) pertains to this section. Most of the Loculatae taxa require SEM to be described properly.

5. Discussion

5.1. South Pacific assemblages and historical records

Previous surveys on marine Achnanthales from French Polynesia were focused on the Society Archipelago (Moorea and Tahiti Islands *in* Ricard 1975, Ricard 1977, Coste & Ricard 1990, Table 3). One *Campyloneis*, fifteen *Cocconeis* and twelve *Achnanthes* taxa were listed (Table 3). *Cocconeis vairaensis* Ricard was not observed during our 2010–2020 survey, while several taxa of the same group

Table 3. List of marine Achnanthales from the French Polynesia *in* Ricard (1975), Ricard (1977) and Coste & Ricard (1990). See remarks in text.
*close to a river discharge.

Taxon
Achnanthes brevipes Agardh
A. brevipes var. *angustata* (Greville) Cleve
A. brevipes var. *parvula* (Kützing) Cleve
A. inflata (Kützing) Grunow
A. inflata var. *elata* (Leuduger-Fortmorel) Gandhi
**A. kuwaitensis* Hendey
A. lanceolata (Brebisson) Grunow
A. longipes Agardh
A. östrupi (Cleve-Euler) Hustedt
A. plönensis Hustedt
A. rhombica Østrup
Campyloneis grevillei (Smith) Grunow & Eulenstein
Cocconeis britannica Naegeli
C. cyclophora var. *decora* Schmidt
C. dirupta var. *dirupta*
C. dirupta var. *flexella* (Janisch & Rabenhorst) Grunow
C. distans Gregory
C. heteroidea Hantzsch var. *heteroidea*
C. heteroidea var. *curvirotunda* (Tempère & Brun) Cleve
C. pellucida Hantzsch
C. placentula Ehrenberg var. *placentula*
C. placentula var. *euglypta* (Ehrenberg) Grunow
C. pseudomarginata Gregory var. *pseudomarginata*
C. pseudomarginata var. *intermedia* Grunow
C. scutellum Ehrenberg var. *scutellum*
C. vairaensis Ricard
Schizostauron citronella (Mann) Górecka, Riaux-Gobin, Witkowski
Xenococconeis neocaledonica (Maillard *ex* Lange-Bertalot & Steindorf) Riaux-Gobin, Le Cohu & Romero

were present, of which *C. suzukii* Riaux-Gobin, Compère, Coste, Straub & Taxböck had a slightly lower SV stria density than observed in *C. vairaensis*. See Riaux-Gobin et al. (2014a) for a provisional key discriminating the taxa of this group. Following Coste & Ricard (1990), *Achnanthes kuwaitensis* Hendey and *Xenococconeis neocaledonica* (Maillard ex Lange-Bertalot & Steindorf) Riaux-Gobin, Le Cohu & Romero are halophilous taxa. Among the 29 taxa listed in Table 3, only 8 are in common with those of the 2010–2020 survey (Table 2), considerably enlarging the taxonomic potential of the high islands from the Society.

Benthic diatoms from Guam (Micronesia, Western Pacific) and other nearby islands were studied by Navarro & Lobban (2009), Lobban & Jordan (2010), Lobban et al. 2012, Lobban (2015) and Park et al. (2018), with the following taxa recorded: *Achnanthes brevipes* Agardh, *Schizostauron citronella* (Mann) Górecka, Riaux-Gobin, Witkowski (Górecka et al. 2021), *Achnanthes cuneata* (Grunow) Grunow, *A. longipes* Agardh, *Planothidium campechianum* (Hustedt) Witkowski, Lange-Bertalot & Metzeltin, *Anorthoneis vortex* Sterrenburg (1987), *C. convexa* Giffen, *Cocconeis coronatoides*, *C. dirupta* Gregory (1857), *C. dirupta* var. *flexella* (Janisch & Rabenhorst) Grunow *in* Van Heurck, *C. distans* Gregory, *C. heteroidea* Hantzsch, *C. scutellum*, and *C. scutellum* var. *ornata* (see above comments about the latter determination that refers to a species pertaining to the Loculatae Section). Lobban (2015, figs 27–32) determined a taxon as *C. subtilissima* Meister, although it is probably closer to *C. suzukii* Riaux-Gobin, Compère, Coste, Straub & Taxböck or to *C. vairaensis* Ricard (see type material examination of *C. subtilissima in* Riaux-Gobin et al. 2014a). The Guam assemblage, concerning the genus *Cocconeis*, is less diversified than reported from the present French Polynesia survey (Table 2), but several taxa were common to both localities.

Diatom assemblages from Galápagos (Equatorial East Pacific) were studied, e.g. by Hendey (1971), Stidolph et al. (2012), Seddon et al. (2014) and Łopato (2016, LM, with ca. sixteen Achnanthales, in part undetermined; the latter material will be accurately re-examined with SEM by Kryk et al., University of Szczecin, Poland, pers. comm.). Surprisingly, of the numerous Achnanthales listed by Hendey (1971) and Stidolph et al. (2012) (32 taxa, Table 4), only *Cocconeis scutellum* and *C. dirupta* var. *flexella* were common to both lists. This remark motivates further studies to focus on this seemingly rich order in the Galápagos, particularly concerning *Achnanthes* sensu lato and *Schizostauron* Grunow.

Navarro (2002) studied marine diatoms from Easter Island, with the description of the new taxon *Florella pascuensis* Navarro, but, to our knowledge, no Achnanthales were studied from Easter islands, except from the freshwater environment of three volcanic crater-lakes (Cocquyt 1991, with six *Achnanthes* species and *Cocconeis placentula*, *C. placentula* var. *euglypta* and *C. scutellum*).

A recent monograph of benthic marine diatoms from the Revillagigedo Archipelago, Mexico (Siquieiros-Beltrones et al. 2021, LM), listed four *Achnanthes*, ten *Cocconeis*, one *Planothidium* and two *Amphicocconeis* taxa. Only based on LM observation (ref. cit.), several taxa would merit a re-investigation,

Table 4. Achnanthales reported from Galápagos by Hendey 1971 and Stidolph et al. 2012.

Taxon	Hendey 1971	Stidolph et al. 2012
Achnanthes angustata Greville	+	–
Achnanthes apiculata Riaux-Gobin, Compère, Hinz & Ector	–	+
Achnanthes brevipes Agardh var. *brevipes*	+	–
Achnanthes brevipes var. *minor* Peragallo	+	–
Achnanthes groenlandica Cleve	+	–
Achnanthes hauckiana Grunow var. *hauckiana*	+	–
Achnanthes hauckiana var. *rostrata* Schultz	+	–
Achnanthes longipes Agardh	–	+
Achnanthes manifera Brun	–	+
Achnanthes parvula Kützing	+	–
Achnanthes sp.	–	+
Amphicocconeis debesi (Hustedt) *in* De Stefano et al.	–	+
Cocconeis costata Gregory	+	–
Cocconeis dirupta Gregory var. *dirupta*	–	+
Cocconeis dirupta var. *flexella* (Janisch & Rabenhorst) Grunow	+	+
Cocconeis distans Gregory	+	–
Cocconeis grata Schmidt	+	–
Cocconeis heteroidea Hantzsch	+	–
Cocconeis heteroidea var. *curvirotunda* (Tempère & Brun) Cleve	–	+
Cocconeis molesta Kützing	+	–

Cocconeis peltoides Hustedt	+	–
Cocconeis placentula var. lineata (Ehrenberg) Van Heurck	–	+
Cocconeis pseudomarginata Gregory	–	+
Cocconeis reticulata Cleve	+	–
Cocconeis scutellum Ehrenberg var. scutellum	+	+
Cocconeis scutellum (= var. adjuncta in Peragallo & Peragallo)	–	+
Cocconeis scutellum var. ornata Grunow	+	–
Cocconeis scutellum var. parva Grunow in Cleve	+	–
Cocconeis stauroneiformis (Van Heurck) Okuno	+	–
Cocconeis sp.	–	+
Schizostauron citronella (Mann) Górecka, Riaux-Gobin, Witkowski	–	+
Vikingea gibbocalyx (Brun) Witkowski, Lange-Bertalot & Metzeltin	+	–
Vikingea promunturi (Giffen) Witkowski, Lange-Bertalot & Metzeltin	+	–

i.e. *Cocconeis hauniensis* (ref. cit., fig. 14h illustrates a taxon larger than *C. hauniensis*, see original description *in* Witkowski et al. 2000, also illustrated *in* Riaux-Gobin & Romero 2003, Riaux-Gobin et al. 2011a) and the *Amphicocconeis* spp. [i.e. *A. disculoides* (Hustedt) De Stefano & Marino 2003 and *A. debesii* (Hustedt) De Stefano 2006] from which SEM illustrations of both valves and valvocopulae are needed before effective classification can be made. *Cocconeis caribensis* Romero & Navarro (synonym *C. paniformis* Brun) has an RV fascia that differentiates it from *C. tuamotuana* (Riaux-Gobin et al. 2015b), a taxon pertaining to the same group, but with no fascia and different stria densities. The latter character cannot be checked in their fig. 12a (ref. cit.). *Cocconeis comis* Schmidt, in their figs 12b, c (ref. cit.), may represent *C. napukensis* Riaux-Gobin, Compère, Straub & Taxböck or *C. meisteri* Riaux-Gobin, Compère, Straub & Taxböck (Riaux-Gobin et al. 2014a). Probably, their fig. 14 l, m (ref. cit.) is not *C. placentula* but a taxon close to *C. coronatoides*. SEM observations would shed more light on the latter interesting assemblage close to the Mexican coast, allowing a better comparison with the taxa from the Galápagos (Table 4). Given the taxonomic uncertainties above, it would be too risky at present to compare the assemblage lists of the Galápagos and Revillagigedo with those previously cited here from the Central Pacific.

5.2. Assemblages from other tropical basins

5.2.1. Tropical Indian Ocean

Several Achnanthales present in the tropical Indian Ocean, particularly off Mascarenes (Riaux-Gobin et al. 2011b, Al-Handal et al. 2016), Madagascar (Kryk et al. 2020) and Scattered Islands (Riaux-Gobin et al. 2018c), are not recorded in Polynesia, e.g. *Achnanthes pulchella* Heiden, *Achnanthidium pseudodelicatissimum* Riaux-Gobin, Compère & Witkowski, *Amphicocconeis rodriguensis* Riaux-Gobin & Al-Handal, *A. debesi*, *C. dirupta* var. *flexella*, *C. inaequalistriata* Riaux-Gobin, Romero, Compère & Al-Handal, *C. krammeri* Lange-Bertalot & Metzeltin (only one possible RV was observed in the Pacific, so it is not mentioned in Table 2), *C.* cf. *sovereinii* Hustedt, *Planothidium delicatulum* (Kützing) Round & Bukhtiyarova, and *P. juandenovense* Riaux-Gobin & Witkowski. *Vikingea florifera* Riaux-Gobin, Romero, Compère & Al-Handal is present in Melanesia, New Caledonia (A.W. pers. obs.) but was not observed in French Polynesia. In contrast, among the cited taxa from Polynesia (Table 2), 21 taxa were not observed in the Mascarenes (Riaux-Gobin et al. 2011c).

As a first attempt, a Venn Diagram (R core Team 2018, Fig. 3, Table 2) was created to illustrate the dissimilarities within assemblages from the Indo-Pacific Basin [Central Polynesia (French Polynesia 2010–2020 survey), Melanesia (New Caledonia 2019 survey) and Indian Ocean (Mascarenes 2005–2009 survey)], and indicated 20 taxa unique to French Polynesia (# in Table 2), seven to New Caledonia and eight to Mascarenes. 26 taxa were common to the three sectors or basins (Fig. 3), suggesting a pantropical distribution for several of them, or a cosmopolitan distribution for taxa such as *C. scutellum* var. *scutel-*

lum. About the taxa unique to each group of islands or basin (Fig. 3), they may qualify as 'potential endemics', even if this concept is still a matter of debate (see '*Marine eukaryote endemism, myth or reality*').

5.2.2. West Atlantic (Florida and Caribbean Arc)

From the West Atlantic, assemblages from Florida, Cuba and the Gulf of Mexico (cf. Navarro 1982, Foged 1984, Navarro & Hernández-Becerril 1997, Frankovich & Wachnicka 2015) are listed in Table 5. The assemblage from the Florida Keys (Montgomery 1978), is commented on separately. Desrosiers et al. (2014, Martinique) created the new genus *Madinithidium* Desrosiers, Witkowski & Riaux-Gobin, with the new species *Madinithidium undulatum* Desrosiers & Witkowski and renamed several *Achnanthidium* Kützing [i.e. *Madinithidium capitatum* (Riaux-Gobin, Romero, Compère & Al-Handal) Witkowski, Riaux-Gobin & Desrosiers, *M. flexuistriatum* (Riaux-Gobin, Compère & A.Witkowski) Witkowski, Riaux-Gobin & Desrosiers, *M. pseudodelicatissimum* (Riaux-Gobin, Compère & Witkowski) Witkowski, Riaux-Gobin & Desrosiers and *M. scalariforme* (Riaux-Gobin, Compère & Witkowski) Witkowski, Riaux-Gobin & Desrosiers]. From the latter *Madinithidium* taxa, only *M. capitatum*, *M. flexuistriatum* and *M. scalariforme* were observed during our 2010–2020 Pacific survey.

Of the 20 taxa listed from the Florida Bay (Frankovich & Wachnicka 2015), thirteen were absent from the assemblages listed by Foged (1984) and Navarro (1982). Among the three latter studies, only four taxa were in common (*Cocconeis britannica* Naegeli ex Kützing, *C. scutellum* var. *scutellum*, *C. euglypta* and *Schizostauron citronella*), and the assemblage from Cuba had a high species richness, with several *Achnanthes* taxa that were absent from the other sites. The assemblage listed by Navarro & Hernández-Becerril 1997 (Caribbean Sea, Mexico, Table 5) was highly diverse with 23 taxa not recorded in the three other monographs.

From the Florida Keys, Montgomery (1978) illustrated with the SEM, numerous Achnanthales (ca. 60), most of them with their SV or RV, but rarely with both valves. Some of these determinations need to be re-examined, even at the genus level. i.e. pl. 60F (ref. cit.) represents *Cocconeis carinata* Riaux-Gobin, Ector & Witkowski as *C.* sp. 5; pl. 60E-F (ref. cit.) represents *C. coronatoides* as *C.* sp. 4; pl. 62A (ref. cit.) may be *C. peltoides* var. *archaeana* Riaux-Gobin & Compère, and pl. 60 (ref. cit.) may be *C. peltoides* [in place of *C. disculus* (Schumann) Cleve]. A careful re-check of this excessively rich material would help to circumscribe the exact Florida assemblage. Numerous taxa listed in Montgomery (1978) are not cited in the four other studies from the West Atlantic (Table 5), increasing to almost 87 the number of Achnanthales taxa present in this oceanic sector, with a large representation by the genus *Achnanthes* (26 taxa).

Nevertheless, it is difficult to establish an objective comparison between the Achnanthales assemblages from the Indo-Pacific Basin versus Tropical West Atlantic sector, only based on the historical monographs reported here, since the sampling and taxonomic effort (via LM or SEM observation), as well as the

Table 5. List of Achnanthales reported from West Atlantic by Navarro 1982 (Florida, Indian River)[2], Foged 1984 (Cuba)[3], Navarro & Hernández-Becerril (1997, Caribbean Sea, Mexico)[4], Frankovich & Wachnicka 2015 (Florida Bay)[1]. *see *Achnanthes* or *Schizostauron* (see intricate history *in* Riaux-Gobin et al. 2015a).

Taxon	Florida Bay[1]	Florida (Indian River)[2]	Cuba[3]	Caribbean[4]
Achnanthes apiculata (Greville) Riaux-Gobin, Compère, Hinz & Ector	–	–	+	–
Achnanthes bengalensis Grunow	–	–	–	+
Achnanthes brevipes Agardh	–	–	–	+
Achnanthes brevipes var. *angustata* (Greville) Cleve	–	+	+	+
Achnanthes brevipes var. *indica* (Brun) Cleve	–	–	–	+
Achnanthes brevipes var. *intermedia* (Kützing) Cleve	–	–	+	+
Achnanthes brevipes var. *parvula* (Kützing) Cleve	–	+	+	+
Achnanthes curvirostrum Brun	–	–	+	+
Achnanthes delicatula (Kütz.) Grunow	–	–	–	+
Achnanthes exigua Grunow	–	–	+	–
Achnanthes exigua var. *heterovalvata* Krasske	–	–	+	+
Achnanthes hauckiana Grunow	–	–	+	–
Achnanthes inflata (Kützing) Grunow	–	–	–	+
Achnanthes javanica Grunow	–	–	–	–
Achnanthes kuwaitensis Hendey	–	+	–	–
Achnanthes lanceolata (Brébisson) Grunow	–	–	+	–
Achnanthes longipes Agardh	–	–	–	+
Achnanthes linearis (Smith) Grunow	–	–	+	–

Taxon				
Achnanthes longipes Agardh, nom. illeg.	−	−	+	−
Achnanthes longipes fo. elliptica Foged	+	−	+	+
Achnanthes manifera Brun	−	+	+	+
Achnanthes minutissima Kützing	−	−	+	−
Achnanthes minutissima var. cryptocephala Grunow	−	−	−	+
Achnanthes perminuta Østrup	−	+	+	−
Amphicocconeis disculoides (Hustedt) De Stefano & Marino	−	−	+	−
Anorthoneis eurystoma Cleve	−	−	+	+
Anorthoneis hyalina Hustedt	−	−	+	−
Anorthoneis japonica Cleve	−	−	+	−
Campyloneis argus Grunow	−	−	+	+
*Cocconeis apiculata Schmidt	−	−	−	−
Cocconeis barleyi Frankovich & De Stefano	+	−	−	−
Cocconeis clandestina Schmidt	+	−	−	+
Cocconeis britannica Naegeli in Kützing	+	+	+	+
Cocconeis convexa Giffen	−	+	−	−
Cocconeis coralliensis Riaux-Gobin & Compère	+	−	+	+
Cocconeis dirupta Gregory	+	−	+	+
Cocconeis dirupta var. flexella (Jan. & Rab.) Grunow	−	−	−	+
Cocconeis discrepans Schmidt	+	−	−	+
Cocconeis disculoides Hustedt	−	−	−	+
Cocconeis distans Gregory	−	−	+	+

Taxon	Florida Bay[1]	Florida (Indian River)[2]	Cuba[3]	Caribbean[4]
Cocconeis distantula Giffen	+	–	–	–
Cocconeis euglypta Ehrenberg	+	+	+	–
Cocconeis fasciolata (Ehr.) Brown	–	–	–	+
Cocconeis fraudulens Simonsen		–	–	–
Cocconeis heteroidea Hantzsch	–	+	+	–
Cocconeis finmarchica Grunow	–	–	–	+
Cocconeis grata Schmidt	–	–	–	+
Cocconeis heteroidea var. *conspicua* (Schmidt) Cleve	–	–	–	+
Cocconeis inconspicua Greville	–	–	–	+
Cocconeis lineata Ehrenberg	+	–	–	–
Cocconeis lyra Schmidt	–	–	–	+
Cocconeis marginifera Østrup	–	–	–	+
Cocconeis maxima (Grunow) Peragallo & Peragallo	+	–	+	+
Cocconeis pediculus Ehrenberg	+	–	+	+
Cocconeis pellucida Grunow	–	+	+	+
Cocconeis pellucida var. *minor* Grunow	–	–	–	+
Cocconeis pelta Schmidt	–	–	–	+
Cocconeis placentula Ehrenberg	–	–	+	+
Cocconeis placentula var. *euglypta* (Ehr.) Grunow	–	–	–	+
Cocconeis placentula var. *lineata* (Ehrenberg) Cleve	–	–	+	–

Species				
Cocconeis pseudodiruptoides Foged	+	–	–	–
Cocconeis scutellum Ehrenberg var. baldjikiana (Grunow 1888) Cleve	+	–	–	–
Cocconeis scutellum Ehrenberg var. parva (Grunow) Cleve	+	–	–	+
Cocconeis scutellum Ehrenberg var. scutellum	+	+	+	+
Cocconeis scutellum var. parva Grunow	–	–	+	–
Cocconeis singularis Hagelstein	–	–	–	+
Cocconeis stauroneiformis (Smith) Okuno	+	+	+	+
Cocconeis thalassiana Romero & López-Fuerte	+	–	–	–
Cocconeis vitrea Brun	+	+	–	–
Cocconeis woodii Reyes-Vasquez	+	+	+	–
Planothidium campechianum (Hustedt) Witkowski & Lange-Bertalot	+	–	–	–
Planothidium cf. pericavum (Carter) Lange-Bertalot	+	–	–	–
Schizostauron citronella (Mann) Górecka, Riaux-Gobin, Witkowski	+	+	+	+

bibliographic references used for the determinations, greatly differed among the studies.

5.3. Marine eukaryote endemism, myth or reality

While some authors still support the 'everything is everywhere' theory concerning eukaryotes (e.g. Finlay 2002), we follow several authors (e.g. Sabbe et al. 2001, Vyverman et al. 2010, Poulíčková et al. 2010, Mann & Vanormelingen 2013, Verleyen et al. 2021) who assume that 'local endemism' may still occur, such as the narrow Antarctic endemism concerning fresh water diatoms (see cit. refs). A first step would be to ascertain a marine large-scale endemism, with pan-tropical taxa versus temperate and polar ones. The presence of taxa only linked to one oceanic sector, and distinctive 'morphs' or 'cryptic taxa', will possibly ascertain a 'potential endemism' at a narrower scale, even for marine benthic diatoms. Genetics will refine or refute these assumptions when cultures of these diatoms, followed by genetic analyses, will be fully possible and results available e.g. via GenBank. Up to now, most of the recent advances in the field of 'genetics versus morphology' concerned freshwater diatoms (i.e. Kermarrec et al. 2013, 2014; Zimmermann et al. 2015; Dulias et al. 2017; Rivera et al. 2018; Mora et al. 2019) while only a few studies concerned marine environments (i.e. An et al. 2017, 2018; Kryk et al. 2020). A recent study on marine Mediterranean biotopes, combining morphological (LM) and DNA metabarcoding (Pérez-Burillo et al. 2022), suggested 'that the way forward, for the moment, is to develop metabarcoding and morphological approaches in parallel and exploit their particular strengths and complementarity'. Another recent study, on freshwater environments (Borrego-Ramos et al. 2021), follows the same sense. Nevertheless, in our opinion, using only LM observations is insufficient and needs to be complemented by those from the SEM, even if this is a time-consuming and expensive method, and by implementing cultures, particularly about small marine benthic taxa, to supplement genetic databases.

From the present tropical survey, only based on ultrastructure, several newly discovered taxa are regarded as 'potential endemics' since they are only found in one sector of the Indo-Pacific Basin (see Venn diagram, Fig. 3). Future genetic surveys will, or will not, validate these assumptions.

The Indo-Pacific Basin is composed of two large interconnected units, while the tropical Atlantic has no direct connection with the two latter entities. The floras of these three oceanic units can thus be compared, taking into account the restrictive remarks i.e. concerning the different methodologies used in these studies.

As a first attempt, and following historical surveys, several taxa from the tropical West Atlantic are, to our knowledge, absent from the tropical Indo-Pacific Basin (i.e. comparison can be made between assemblages cited in Montgomery 1978 and Table 5, versus Table 2).

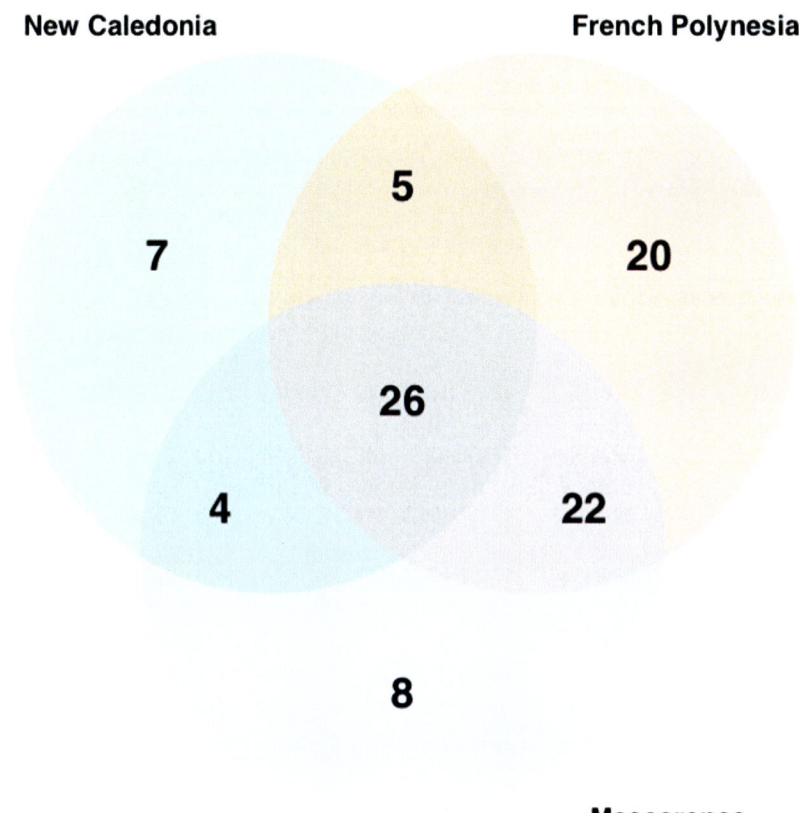

Fig. 3. Venn Diagram (R core Team 2018). Dissimilarities within Achnanthales assemblages from French Polynesia, New Caledonia and Mascarenes, with 20 taxa unique to French Polynesia (# *in* Table 2).

5.4. Role of insularity, position in the SEC and geologic past

In the Tuamotu, with regards to the low diatom colonization of atolls (e.g. Napuka, Takaroa, Takapoto, Nukutavake, C.R.-G. pers. obs.), the percentage of new or rarely mentioned *Cocconeis* is high: Among the eleven Achnanthales taxa reported from Napuka, three were described as new (*Cocconeis frustrationis*, *C. tuamotuana* and *C. napukensis*) and a fourth taxon has been, more recently, described as new from a saline lake in Iraq (*Cocconeis sawensis* Al-Handal & Riaux-Gobin, Al-Handal et al. 2014). *Cocconeis sawensis* is thus a rare pantropical taxon present in diverse environments characterized by high salinity.

Table 6. Location of the Types (South Pacific 2010–2020 surveys, Table 1–2): BM numbers = in NHM diatom collection (National Museum London).
* *in* Riaux-Gobin collection (CRIOBE-UPVD, Perpignan, France)
** *in* Witkowski collection (Szczecin University, Szczecin, Poland)
*** *in* Hustedt collection (Alfred-Wegener-Institute für Polar-und Meeresforschung, Bremerhaven, Germany)
For illustration of the Holotype specimens, see corresponding publication. Year refers to the sampling.

Sample/Names	Location	GPS geo-localization	Year	Holotype/Isotypes (slide or stub)
Society Archipelago				
COOK1	Moorea	17°30.431'S, 149°49.222'W	2010	1 10/06/2011*
MOOR1	Moorea	17°31.047'S, 149°51.021'W	2010	2 21/01/2011*
Papeete4 Holoth	Tahiti-Nui	17°31.405'S, 149°31.106'W	2010	2 13/01/2012*
Papeete4 Holoth	Tahiti-Nui	17°31.405'S, 149°31.106'W	2010	BM001222888
Papeete4 Holoth	Tahiti-Nui	17°31.405'S, 149°31.106'W	2010	BM001222885
Tuamotu Archipelago				
15NPK7-2	Napuka	14°10.702'S, 141°15.949'W	2012	1 13/12/2013*
15NPK7-2	Napuka	14°10.702'S, 141°15.949'W	2012	5 28/03/2014*
10NPK3-1	Napuka	14°10.454'S, 141°16.315'W	2012	6 14/12/2012*
NTVA1	Nukutavake	19°16.833'S, 138°47.116'W	2016	BM101912
Austral Archipelago				
Rapa-1	Rapa	27°28'S, 144°20'W	2014	BM001222256
RAI20	Raivavae	23°51.274'S, 147°39.595'W	2018	BM81917
RAI20	Raivavae	23°51.274'S, 147°39.595'W	2018	BM001222887
Marquesas Archipelago				
NH4-2	Nuku Hiva	08°55.140'S, 140°05.600'W	2015	BM101950

Syntypes	Species	Reference
BM101626, ZU8/15*** MOOR1*	Cocconeis spina-christi	Riaux-Gobin et al. 2013a
BM101629, ZU8/31*** MOOR2	Xenococconeis opunohusiensis	Riaux-Gobin et al. 2014
BM101645, ZU8/38*** TAHITI1*	Astartiella societatis	Riaux-Gobin et al. 2013b
BM101645, ZU8/38***	Amphicocconeis clypeus	Riaux-Gobin et al. 2021c
BM001222886, BM101645	Amphicocconeis antiqua	Riaux-Gobin et al. 2021c
BM101713, NPK1*	Cocconeis frustrationis	Riaux-Gobin et al. 2015b
BM101713, NPK1*	Cocconeis tuamotuana	Riaux-Gobin et al. 2015b
BM101782, NPK2*	Pseudachnanthidium megapteropsis	Riaux-Gobin & Witkowski 2015
NTVA1*, SZCZ24377**	Cocconeis kurakakea	Riaux-Gobin et al. 2018c
	Cocconeis santandrea	Riaux-Gobin et al. 2021b
	Cocconeis vaiamanuensis	Riaux-Gobin et al. 2021d
BM81917	Amphicocconeis ruatara	Riaux-Gobin et al. 2021c
	Cocconeis carinata	Riaux-Gobin et al. 2019a

In the volcanic environment of high islands, e.g. Moorea and Tahiti Islands (South Polynesia, Society), the Achnanthales species richness (>40 taxa) is on average three times higher than in atolls from Tuamotu, with dense quantitative populations, but the occurence of new or rare species is proportionally quite low. In the main part, the assemblages of high islands are composed of ubiquitous and pantropical taxa of which several are in common with those from the Indian Ocean (i.e. Mascarenes, cf. Riaux-Gobin et al. 2011b).

5.5. Concluding remarks

The present Achnanthales overview is a first attempt, that will have to be improved upon in the future, particularly concerning *Planothidium* and *Achnanthes* taxa, as the present study focusses more in-depth on the Cocconeidaceae, notably on rare and small taxa from French Polynesia, which is an echo to a previous study of the Mascarenes using the same methodologies.

The taxa from the tropical Indo-Pacific Basin are largely ubiquitous pantropical, whereas a few taxa are only present at one pole of the basin. Some taxa are restricted to the north of Polynesia (i.e. Tuamotu), with a possible correlation to the geologic past of the archipelago and position of islands in the SEC, allowing a higher biogeographic insulation.

Among the large Indo-Pacific Basin, the two oceanic entities (tropical sectors of the Pacific Ocean and Indian Ocean) each seem to have some slight specificities i.e. some taxa are related to only one basin. A stronger comparison with the tropical West Atlantic assemblages would refine our morphological analysis and permit a more accurate notion of endemism for marine eukaryotes. The greater contribution by genetics will certainly provide valuable assistance in supporting or refuting ultrastructural analyzes.

Acknowledgments: Jean Kape 'Fasan Chong', Remy Tchong and Marina Taki are warmly acknowledged for their assistance during our stay in Napuka atoll, Tomy Ung for his hospitality and accomplice friendship during our Takaroa samplings, Xavier Curvat 'Pipapo' for collecting epizoic samples by diving in Marquesas. Thanks are also due to René Galzin (CNRS–UAR 3278CRIOBE, Perpignan, France) for his logistic help and friendship during sampling in Tahiti-iti, and to our late colleague Luc Ector (Luxembourg Institute of Science and Technology (LIST), Belvaux, Luxembourg) for invaluable bibliographic help. A particular acknowledgment to Wim Vyverman (Protistology and Aquatic Ecology, Department of Biology, Ghent University (UGent), 9000 Ghent, Belgium) for his discussions about endemism, and to Pablo Saenz-Agudelo (Universidad Austral de Chile, Instituto de Ciencias Ambientales y Evolutivas Valdivia, Region de Los Rios, Chile) for building the Venn diagram. Mitko Gorand and afterwards Yonko Gorand (C2M, University of Perpignan-Via Domitia) are acknowledged for their kind SEM assistance during the last 25 years. We acknowledge the CNRS–UAR 3278 for supporting this research

6. References

Al-Handal, A.Y., Riaux-Gobin, C., Abdulla, D.S. & Ali, M.H. (2014). *Cocconeis sawensis* sp. nov. (Bacillariophyceae) from a saline lake (Sawa Lake), South Iraq: comparison with allied taxa. *Phytotaxa, 181*(4), 216–228.

Al-Handal, A.Y., Compère, P. & Riaux-Gobin, C. (2016). Marine benthic diatoms in the coral reefs of Reunion and Rodrigues Islands, West Indian Ocean. *Micronesica, 2016*(3), 1–78.

Al-Handal, A.Y., Romero, O.E., Eggers, S.L. & Wulff, A. (2021). *Navithidium* gen. nov., a new monoraphid diatom (Bacillariophyceae) genus based on *Achnanthes delicatissima* Simonsen. *Diatom Research, 36*(2), 133–141. https://doi.org/10.1080/0269249X.2021.1921039

An, S.M., Choi, D.H., Lee, J.H., Lee, H. & Noh, J.H. (2017). Identification of benthic diatoms isolated from the eastern tidal flats of the Yellow Sea: Comparison between morphological and molecular approaches. *PLoS ONE, 12*(6), e0179422. https://doi.org/10.1371/journal.pone.0179422

An, S.M., Choi, D.H., Lee, J.H., Lee, H. & Noh, J.H. (2018). Next-generation sequencing reveals the diversity of benthic diatoms in tidal flats. *Algae, 33*(2), 167–180. https://doi.org/10.4490/algae.2018.33.4.3

Borrego-Ramos, M., Bécares, E., García, P., Nistal, A. & Blanco, S. (2021). Epiphytic Diatom-Based Biomonitoring in Mediterranean Ponds: Traditional Microscopy versus Metabarcoding Approaches. *Water, 13*, 1351. https://doi.org/10.3390/w13101351

Bory de Saint-Vincent, J.B.G.M. (1822). *Achnanthe. Achnanthes.* In I. Audouin et al. (eds), Dictionnaire Classique d'Histoire Naturelle. Vol. 1, pp. 79–80. Paris.

Bouniot, E. & Jourdan, S. (2009). *A Pacific tropical islands typology using GIS*. Initially published in French for the 2009 Pacific Science Meeting in Tahiti. https://www.psi2009.pf/

Brousse, R., Barsczus, H.G., Bellon, H., Cantagrel, J.M., Diraison, C., Guillou, H. & Leotot, C. (1990). The Marquesas alignment (French-Polynesia)-Volcanology, geochronology, a hot-spot model. *Bulletin de la société Géologique de France, 6*(6), 933–949.

Car, A., Witkowski, A., Dobosz, S., Burfeind, D.D., Meinesz, A., Jasprica, N., Ruppel, M., Kurzydlowski, K.J. & Plocinski, T. (2012). Description of a new marine diatom, *Cocconeis caulerpacola* sp. nov. (Bacillariophyceae), epiphytic on invasive *Caulerpa* species. *European Journal of Phycology, 47*(4), 433–448.

Cavaloc, E. (1988). *Colonisation des Rhizophora (Palétuviers) récemment introduits à Moorea (Société, Polynésie Française). Bilan de répartition (1987) et conséquences écologiques.* EPHE, Rapport RA 28, 43 pp.

Chauvel, C., Maury, R.C., Blais, S., Lewin, E., Guillou, H., Guille, H., Rossi, P. & Gutscher, M.-A. (2012). The size of plume heterogeneities constrained by Marquesas isotopic stripes. *Geochemistry, Geophysics, Geosystems, 13*(7), Q07005. https://doi.org/10.1029/2012GC004123

Cleve, P.T. & Grunow, A. (1880). Beiträge zur Kenntniss der Arktischen Diatomeen *Kongliga Svenska-Vetenskaps Akademiens Handlingar, 17*(2), 121 pp., 7 pls.

Cocquyt, C. (1991). Diatoms from Easter Island. *Biologisch Jaarboek [Dodonaea], 59*, 109–124.

Corlett, H. & Jones, B. (2007). Epiphyte communities on *Thalassia testudinum* from Grand Cayman, British West Indies: Their composition, structure, and contribution to lagoonal sediments. *Sedimentary Geology, 194*(3–4), 245–262.

Costa, L.F., Wetzel, C.E., Ector, L. & Bicudo, D.C. (2020). Freshwater *Cocconeis* species (Bacillariophyceae) from Southeastern Brazil, and description of *C. amerieuglypta* sp. nov. *Botany Letters, 167*(1), 15–31. https://doi.org/10.1080/23818107.2019.1672103

Coste, M. & Ricard, M. (1990). Diatomées continentales des Îles de Tahiti et de Moorea (Polynésie Française). In M. Ricard & M. Coste (eds), *Ouvrage dédié à la Mémoire du Professeur Henry Germain (1903–1989)*. (pp. 33–62). Königstein: Koeltz Scientific Books.

Desianti, N., Potapova, M & Beals, J. (2015). Examination of the type materials of diatoms described by Hohn and Hellerman from the Atlantic Coast of the USA. *Diatom Research, 30*(2), 1–24. https://doi.org/10.1080/0269249X.2014.1000020

Desrosiers, C., Witkowski, A., Riaux-Gobin, C., Zglobicka, I., Kurzydlowski, K.J., Eulin, A., Leflaive, J. & Ten-Hage, L. (2014). *Madinithidium* gen. nov. (Bacillariophyceae), a new monoraphid diatom genus from the tropical marine coastal zone. *Phycologia, 53*(6), 583–592.

De Stefano, M. & Marino, D. (2001). Comparison of *Cocconeis pseudonotata* sp. nov. with two closely related species, *C. notata* and *C. diruptoides*, from *Posidonia oceanica* leaves. *European Journal of Phycology, 36*(4), 295–306. https://doi.org/10.1080/09670260110001735458

De Stefano, M. & Marino, D. (2003). Morphology and taxonomy of *Amphicocconeis* gen. nov. (Achnanthales, Bacillariophyceae, Bacillariophyta) with considerations on its relationship to other monoraphid diatom genera. *European Journal of Phycology, 38*, 361–370. https://doi.org/10.1080/0967026031000612646

De Stefano, M. & Romero, O. (2005). A survey of alveolate species of the diatom genus *Cocconeis* (Ehr.) with remarks on the new section *Alveolatae*. *Bibliotheca Diatomologica, 52*, 1–132.

De Stefano, M., Marino, D. & Mazzella, L. (2000). Marine taxa of *Cocconeis* on leaves of *Posidonia oceanica*, including a new species and two new varieties. *European Journal of Phycology, 35*(3), 225–242. https://doi.org/10.1080/09670260010001735831

De Stefano, M., Romero, O.E. & Totti, C. (2008). A comparative study of *Cocconeis scutellum* Ehrenberg and its varieties (Bacillariophyta). *Botanica Marina, 51*(6), 506–536. https://doi.org/10.1515/bot.2008.058

Diraison, C. (1991). *Le volcanisme aérien des archipels polynésiens de la Société, des Marquises et des Australes-Cook. Téphrostratigraphie, datation isotopique et géochimie comparées. Contribution à l'étude des origines du volcanisme intraplaque du Pacifique central.* (413 pp.) PhD Thesis, Université de Bretagne Occidentale, Brest, France.

Dulias, K., Stoof-Leichsenring, K.R., Pestryakova, L.A. & Herzschuh, U.. (2017). Sedimentary DNA versus morphology in the analysis of diatom-environment relationships. *Journal of Paleolimnology, 57*, 51–66. https://doi.org/10.1007/s10933-016-9926-y

Ehrenberg, C.G. (1836). Zusätze zur Erkenntniss grosser organischer Ausbildung in den kleinsten thierischen Organismen. *Abhandlungen der Königlichen Akademie der Wissenschaften zu Berlin, 1835*, 150–181.

Ehrenberg, C.G. (1838). *Atlas von vier und sechzig Kupfertafeln zu Christian Gottfried Ehrenberg über Infusionsthierchen*. pls I-LXIV. Leipzig: Verlag von Leopold Voss.

Ehrenberg, C.G. (1843). Verbreitung und Einfluss des mikroskopischen Lebens in Süd- und Nord-Amerika. *Abhandlungen der Königlichen Akademie der Wissenschaften zu Berlin, 1841*, 291–445, 4 pls. https://biodiversitylibrary.org/page/29106994

Ehrenberg, C.G. (1849). Passatstaub und Blutregen. Ein großes organisches unsichtbares Wirken und Leben in der Atmosphäre. *Abhandlungen der Königlichen Akademie der Wissenschaften zu Berlin, 1847*, 269–460, 6 pl.

Ehrenberg, C.G. (1854). *Mikrogeologie*. Einundvierzig Tafeln mit über viertausend grossentheils colorirten Figuren, Gezeichnet vom Verfasser. pp. [1]–31, 40 pls [Taf. I–XXXX]. Leipzig: Verlag von Leopold Voss.

Finlay, B.J. (2002). Global dispersal of free-living microbial eukaryote species. *Science, 296*(5570), 1061–1063. https://doi.org/10.1126/science.1070710

Foged, N. (1975). Some littoral diatoms from the coast of Tanzania. *Bibliotheca Phycologica, 16*, 1–127.

Foged, N. (1979). Diatoms in New Zealand, the North Island. *Bibliotheca Phycologica, 47*, 1–225.

Foged, N. (1984). Freshwater and littoral diatoms from Cuba. *Bibliotheca Diatomologica, 5*, 1–243.

Foged, N. (1987). Diatoms from Viti Levu, Fiji Islands. *Bibliotheca Diatomologica, 14*, 1–195.

Frankovich, T.A. & Wachnicka, A. (2015). Epiphytic Diatoms along Phosphorus and Salinity Gradients in Florida Bay (Florida, USA), an Illustrated Guide and Annotated Checklist. In J.A. Entry, A.D. Gottlieb, K. Jayachandran & A. Ogram, *Microbiology of the Everglades Ecosystem* (pp. 241–288).

Giffen, M. H. (1967). Contributions to the diatom flora of South Africa III. Diatoms of the marine littoral regions at Kidd's Beach near east London, Cape Province, South Africa. *Nova Hedwigia, 13*, 245–292.

Goreau, T.F., Goreau, N.I. & Goreau, T.J. (1979). Corals and coral reefs. *Scientific American, 241*, 124–136.

Górecka, E., Ashworth, M.P., Davidovich, N., Davidovich, O., Dąbek, P., Sabir, J.S.M. & Witkowski, A. (2021). Multigene phylogenetic data place monoraphid diatoms *Schizostauron* and *Astartiella* along with other fistula-bearing genera in the Stauroneidaceae. *Journal of Phycology, 57*(5), 1472–1491. https://doi.org/10.1111/jpy.13192.

Gregory, W. (1855). On the post-Tertiary lacustrine sand containing diatomaceous exuviae from Glenshire near Inverary. *Quarterly Journal of Microscopical Science, 3*, 30–43.

Gregory, W. (1857). On new forms of marine Diatomaceae found in the Firth of Clyde and in Loch Fyne. *Transactions of the Royal Society of Edinburgh, 21*, 473–542.

Grunow, A. (1868 '1867'). Algae. In *Reise der österreichischen Fregatte Novara um die Erde in den Jahren 1857, 1858, 1859 unter den Befehlen des Commodore B. von Wüllerstorf-Urbair.* Botanischer Theil. Erster Band. Sporenpflanzen. E. Fenzl et al. (eds), pp. 1–104. Wien [Vienna]: Aus der Kaiserlich Königlichen Hof- und Staatsdruckeri in Commission bei Karl Gerold's Sohn.

Guille, G., Legendre, C., Maury, R.C., Caroff, M., Munschy, M., Blais, S., Chauvel, C., Cotten, J. & Guillou, H. (2002). Les Marquises (Polynésie Française): un archipel intraocéanique atypique. *Géologie de la France, 2*, 5–37.

Hendey, N.I. (1971). Some marine diatoms from the Galápagos Islands. *Nova Hedwigia, 22*(1–2): 371–422.

Hohn, M.H. & Hellerman, J. (1966). New diatoms from the Lewes-Rehoboth Canal, Delaware and Chesapeake Bay area of Baltimore, Maryland. *Transactions of the American Microscopical Society, 85*, 115–130.

Hustedt, F. (1933). Die Kieselalgen Deutschlands, Österreichs und der Schweiz unter Berücksichtigung der übrigen Länder Europas sowie der angrenzenden

Meeresgebiete. In L. Rabenhorst (ed.), *Kryptogamen Flora von Deutschland, Österreich und der Schweiz.* Akademische Verlagsgesellschaft m.b.h. Leipzig. 7(Teil 2, Lief. 3), 321–432, figs 781–880.

Hustedt, F. (1957). Die Diatomeenflora des Flußsystems der Weser im Gebiet der Hansestadt Bremen. *Abhandlungen des Naturwissenschaftlichen Verein zu Bremen, 34*(3), 181–440, 1 pl.

Hustedt, F. (1939). Die Diatomeenflora des Küstengebietes der Nordsee vom Dollart bis zur Elbemündung. I. Die Diatomeenflora in den Sedimenten der unteren Ems sowie auf den Watten in der Leybucht, des Memmert und bei der Insel Juist. *Abhandlungen des Naturwissenschaftlichen Verein zu Bremen, 31*(2/3), 571–677.

Hustedt, F. & Aleem, A.A. (1951). Littoral diatoms from the Salstone, near Plymouth. *Journal of the Marine Biological Association of the United Kingdom, 30*(1), 177–196.

Jahn, R., Kusber, W.-H. & Romero, O.E. (2009). *Cocconeis pediculus* Ehrenberg and *C. placentula* Ehrenberg var. *placentula* (Bacillariophyta): Typification and taxonomy. *Fottea, 9*(2), 275–288.

Jahn, R., Abarca, N., Kusber, W.-H., Skibbe, O., Zimmermann, J. & Mora, D. (2020). Integrative taxonomic description of two new species of the *Cocconeis placentula* group (Bacillariophyceae) from Korea based on unialgal strains. *Algae, 35*(4), 303–324. https://doi.org/10.4490/algae.2020.35.8.1

Kermarrec, L., Franc, A., Rimet, F., Chaumeil, P., Humbert, JF. & Bouchez, A. (2013). Next-generation sequencing to inventory taxonomic diversity in eukaryotic communities: a test for freshwater diatoms. *Molecular Ecology Resources, 13*, 607–619.

Kermarrec, L., Franc, A., Rimet, F., Chaumeil, P., Frigerio, J.-M., Humbert, J.-F. & Bouchez, A. (2014). A next-generation sequencing approach to river biomonitoring using benthic diatoms. *Freshwater Science, 33*, 349–363.

Kryk, A., Bąk, M., Górecka, E., Riaux-Gobin, C., Bemiasa, J., Bemajana, E., Li, C., Dąbek, P., Witkowski, A. (2020). Marine diatom assemblages of the Nosy Be Island coasts, NW Madagascar: species composition and biodiversity using molecular and morphological taxonomy. *Systematics and Biodiversity, 18*(2), 161–180. https://doi.org/10.1080/14772000.2019.1696420

Kützing, F.T. (1844). *Die Kieselschaligen Bacillarien oder Diatomeen.* pp. [i-vii], [1]–152, pls 1–30. Nordhausen: W. Köhne.

Le Cohu, R. (1985). Ultrastructure des diatomées de Nouvelle-Calédonie. Première partie. *Annales de Limnologie, 21*, 3–12.

Le Dez, A., Maury R.C., Vidal, P., Bellon, H., Cotton, J. & Brousse R. (1996). Geology and geochemistry of Nuku Hiva, Marquesas: Temporal trends in a large Polynesian shield volcano. *Bulletin de la société Géologique de France, 167*(2), 197–209.

Legendre, C. (2003). *Pétrogenèse de laves différenciées en contexte intraplaque océanique et hétérogénéité géochimique au niveau du point chaud des Marquises (Polynésie Française): étude des îles de Ua Pou et de Nuku Hiva.* Géochimie. Université de Bretagne occidentale, Brest. HAL Id: tel-00008677. https://tel.archives-ouvertes.fr/file/index/docid/47808/filename/tel-00008677.pdf

Lobban, C.S. (2015). Benthic marine flora of Guam: new records, redescription of *Psammdictyon pustulatum* n. comb., n. stat., and three new species (*Colliculoamphora gabgabensis, Lauderia excentrica,* and *Rhoiconeis pagoensis*). *Micronesica, 2015*(02), 1–49.

Lobban, C.S. & Jordan, R.W. (2010). Diatoms on coral reefs and in tropical marine lakes. In J. P. Smol & E.F. Stoermer (Eds.) *The diatoms: applications for the environmental and earth sciences* (2nd ed.) (pp. 346–356). Cambridge University Press, Cambridge. https://doi.org/10.1017/CBO9780511763175.019

Lobban, L., Schefter, M., Jordan, R.W., Arai, Y., Sasaki, A., Theriot, E.C., Ashworth, M., Ruck, C. & Pennesi, C. (2012). Coral-reef diatoms (Bacillariophyta) from Guam: new records and preliminary checklist, with emphasis on epiphytic species from farmer-fish territories. *Micronesica, 43*, 237–479. https://micronesica.org/sites/default/files/11_lobban_et_al.pdf

Łopato, D. (2016). *Taxonomy and biodiversity of diatom assemblages from the littoral zone of Islands: Isabela, Santa Cruz and San Cristobal, Galápagos Archipelago* Unpublished master's thesis. Szczecin, Poland: University of Szczecin.

Maillard, R. (1978). Contribution à la connaissance des diatomées d'eau douce de la Nouvelle-Calédonie (Océanie). *Cahiers ORSTOM, sér. Hydrobiolie, 12*(2), 143–172.

Mann, D.G., Vanormelingen, P. (2013). An inordinate fondness? The number, distributions, and origins of diatom species. *Journal of Eukaryotic Microbiology, 60*(4), 414–420. https://doi.org/10.1111/jeu.12047

Martinez, E., Ganachaud, A., Lefevre, J. & Maamaatuaiahutapu, K. (2009). Central South Pacific thermocline water circulation from a high-resolution ocean model validated against satellite data: Seasonal variability and El Nino 1997–1998 influence. *Journal of Geophysical Research, 114,* C05012. https://doi.org/10.1029/2008JC004824

Mizuno, M. (1987). Morphological variation of the attached diatom Cocconeis scutellum var. scutellum (Bacillariophyceae). *Journal of Phycology, 23,* 591–597.

Monnier, O., Rimet, F., Bey, M., Chavaux, R. & Ector, L. (2007). Sur l'identité de Cocconeis euglypta Ehrenberg 1854 et C. lineata Ehrenberg 1843 – Une approche par les sources historiques. *Diatomania, 11*, 30–45.

Montaggioni, L. (2015). Naissance et évolution géologique des îles Australes. In B. Salvat, T. Bambridge, D. Tanret, & J. Petit (eds). *Environnement marin des îles Australes, Polynésie française.* Institut Récifs Coralliens Pacifique. CRIOBE. The Pew CharitableTrusts Polynésie française. (pp. 29–39).

Montgomery, R.T. (1978). *Environmental and ecological studies of the diatom communities associated with the coral reefs of the Florida Keys. Vols I & II.* Ph.D. dissertation. Thesis. Florida State University, Tallahassee, FL.

Mora, D., Abarca, N., Proft, S., Grau, J.H., Enke, N., Carmona, J., Skibbe, O., Jahn, R. & Zimmermann, J. (2019). Morphology and metabarcoding: a test with stream diatoms from Mexico highlights the complementarity of identification methods. *Freshwater Science, 38*(3), 448–464. https://doi.org/10.1086/704827

Mora, D., Stancheva, R. & Jahn, R. (2021). *Cocconeis czarneckii* sp. nov. (Bacillariophyta): a new diatom species from Lake Okoboji (Iowa, USA), based on the strain UTEX FD23. *Phycologia, 61*(1), 60–74. https://doi.org/10.1080/00318884.2021.2003684

Moser, G., Lange-Bertalot, H. & Metzeltin, D. (1998). Insel der Endemiten. Geobotanisches Phänomen Neukaledonien. *Bibliotheca Diatomologica, 38*, 1–464.

Navarro, J.N. (1982). Marine diatoms associated with mangrove prop roots in the Indian River, Florida, USA. *Bibliotheca Phycologica, 61*, 1–151.

Navarro, J.N. (2002). *Florella pascuensis* sp. nov., a new marine diatom species from Easter Island (Isla de Pascua), Chile. *Diatom Research, 17*(2), 282–289. https://doi.org/10.1080/0269249X.2002.9705548

Navarro, J.N. & Hernández-Becerril, D.U. (1997). Checklist of marine diatoms from the Caribbean Sea. *Listados florísticos de Mexico, 15*. (48 pp.). Universidad Nacional Autónoma de Mexico.

Navarro, J.N. & Lobban, C.S. (2009). Freshwater and marine diatoms from the Western Pacific islands of Yap and Guam, with notes on some diatoms in damselfish territories. *Diatom Research, 24*(1), 123–157. https://doi.org/10.1080/0269249X.2009.9705787

Okuno, H. (1957). Electron-microscopical study of the fine structure of diatom frustules. XVI. *Botanical Magazine [Shokubutsu-gaku zasshi], 70*(829/830), 216–222.

Pandolfi, J.M., Bradbury, R.H., Sala, E., Hughes, T.P., Bjorndal, K.A., Cooke, R.G., McArdle, D., McClenachan, L., Newman, M.J.H., Paredes, G., Warner, R.R. & Jackson, J.B.C. (2003). Global trajectories of long-term decline of coral reef ecosystems. *Science, 301*, 955–958.

Park, J.S., Lobban, C.S. & Lee, K.W. (2018). Diatoms associated with seaweeds from Moen Island in Chuuk Lagoon, Micronesia. *Phytotaxa, 351*(2), 101–140. https://doi.org/10.11646/phytotaxa.351.2.1

Pérez-Burillo, J., Valoti, G., Witkowski, A., Prado, P., Mann, D.G. & Trobajo, R. (2022). Assessment of marine benthic diatom communities: insights from a combined morphological-metabarcoding approach in Mediterranean shallow coastal waters. *Marine Pollution Bulletin, 174*, 113183. https://doi.org/10.1016/j.marpolbul.2021.113183.

Potapova, M. & Ponader, K.C. (2004). Two common North American diatoms, *Achnanthidium rivulare* sp. nov. and *A. deflexum* (Reimer) Kingston: morphology, ecology and comparison with related species. *Diatom Research, 19*(1), 33–57.

Potapova, M. & Spaulding, S. (2013). *Cocconeis placentula* sensu lato. In *Diatoms of North America*. Retrieved March 02, 2022, from https://diatoms.org/species/cocconeis_placentula

Poulíčková, A., Veselá, J., Neustupa, J. & Skaloud, P. (2010). Pseudocryptic diversity versus cosmopolitanism in diatoms: a case study on *Navicula cryptocephala* Kütz. (Bacillariophyceae) and morphologically similar taxa. *Protist, 161*(3), 353–69. https://doi.org/10.1016/j.protis.2009.12.003.

R Core Team (2018). *R: A Language and Environment for Statistical Computing*. R Foundation for Statistical Computing, Vienna. https://www.R-project.org

Riaux-Gobin, C. & Romero, O. (2003). Marine *Cocconeis* Ehrenberg (Bacillariophyceae) species and related taxa from Kerguelen's Land (Austral Ocean, Indian sector). *Bibliotheca Diatomologica, 47*, 1–189.

Riaux-Gobin, C. & Compère, P. (2008). New *Cocconeis* taxa (Bacillariophyceae) from coral sands off Réunion Island (Western Indian Ocean). *Diatom Research, 23*(1), 129–146.

Riaux-Gobin, C., Romero, O.E., Al-Handal, A.Y. & Compère, P. (2010). Two new *Cocconeis* taxa (Bacillariophyceae) from coral sands off the Mascarenes (Western Indian Ocean) and some related unidentified taxa. *European Journal of Phycology, 45*(3), 278–292. https://doi.org/10.1080/09670260903560076

Riaux-Gobin, C., Compère, P. & Al-Handal, A.Y. (2011a). Species of the *Cocconeis peltoides* group with a marginal row of unusual processes (Mascarenes and Kerguelen Islands, Indian Ocean). *Diatom Research, 26*(4), 325–338.

Riaux-Gobin, C., Romero, O.E., Compère, P. & Al-Handal, A.Y. (2011b). Small-sized Achnanthales (Bacillariophyta) from coral sands off Mascarenes (Western Indian Ocean). *Bibliotheca Diatomologica, 57*, 1–234.

Riaux-Gobin, C, Romero, O.E., Coste, M. & Galzin, R. (2013a). A new *Cocconeis* (Bacillariophyceae) from Moorea Island, Society Archipelago, South Pacific Ocean with distinctive valvocopula morphology and linking system. *Botanica Marina, 56*(4), 339–356.

Riaux-Gobin, C., Witkowski, A. & Romero, O.E. (2013b). An account of *Astartiella* species from tropical areas with a description of *A. societatis* sp. nov. and nomenclatural notes. *Diatom Research, 28*(4), 419–430. https://doi.org/10.1080/0269249X.2013.827590

Riaux-Gobin, C., Compère, P., Coste, M., Straub, F. & Taxböck, L. (2014a). *Cocconeis napukensis* sp. nov. (Bacillariophyceae) from Napuka Atoll (South Pacific) and lectotypification of *Cocconeis subtilissima* Meister. *Fottea, 14*(2), 209–224.

Riaux-Gobin, C., Coste, M., Jordan, R.W., Romero, O.E. & Le Cohu, R. (2014b). *Xenococconeis opunohusiensis* gen. et sp. nov. and *Xenococconeis neocaledonica* comb. nov. (Bacillariophyta) from the tropical South Pacific. *Phycological Research, 62*(3), 153–169.

Riaux-Gobin, C., Compère, P., Hinz, F. & Ector, L. (2015a). *Achnanthes citronella, A. trachyderma* comb. nov. Bacillariophyta) and allied taxa pertaining to the same morphological group. *Phytotaxa, 227*, 101–119. https://doi.org/10.11646/phytotaxa.227.2.1.

Riaux-Gobin, C., Compère, P. & Jordan, R.W. (2015b). *Cocconeis* Ehrenberg assemblage (Bacillariophyceae) from Napuka Atoll (Tuamotu Archipelago, South Pacific) with descriptions of two new taxa. *Diatom Research, 30*(2), 175–196. https://doi.org/10.1080/0269249x.2015.1021839

Riaux-Gobin, C., Witkowski, A., Compère, P. & Romero, O.E. (2015c). *Cocconeis* Ehrenberg taxa (Bacillariophyta) with a marginal row of simple processes: relationship with the valvocopula system and distinctive features of related taxa. *Fottea, 15*(2), 139–154. https://doi.org/10.5507/fot.2015.015

Riaux-Gobin, C. & Witkowski, A. (2015d). *Pseudachnanthidium megapteropsis* gen. nov. and sp. nov. (Bacillariophyta): a widespread Indo-Pacific elusive taxon. *Cryptogamie Algologie, 36*(3), 291–304.

Riaux-Gobin, C., Witkowski, A., Chevallier, D. & Daniszewska–Kowalczyk, G. (2017a). Two new *Tursiocola* species (Bacillariophyta) epizoic on green turtles (*Chelonia mydas*) in French Guiana and Eastern Caribbean. *Fottea, 17*, 150–163. https://doi.org/10.5507/fot.2017.007

Riaux-Gobin, C., Witkowski, A., Kociolek, J.P., Ector, L., Chevallier, D. & Compère, P. (2017b). New epizoic diatom (Bacillariophyta) species from sea turtles in the Eastern Caribbean and South Pacific. *Diatom Research, 32*(1), 109–125.

Riaux-Gobin, C., Ector, L., Witkowski, A. & Igersheim, A. (2018a). Achnanthales from historical Grunow collection in Porto Subzanski, Croatia. *Botanica Marina, 61*(6), 573–593. https://doi.org/10.1515/bot-2018-0045

Riaux-Gobin, C., Witkowski, A. & Igersheim, A. (2018b). *Cocconeis scutellum* var. *parva* (Bacillariophyceae) re-examination and typification. *Phytotaxa, 343*(1), 20–34. https://doi.org/10.11646/phytotaxa.343.1.2

Riaux-Gobin, C., Witkowski, A., Igersheim, A., Lobban, C.S., Al-Handal, A.Y. & Compère, P. (2018c). *Planothidium juandenovense* sp. nov. (Bacillariophyta) from Juan de Nova (Scattered Islands, Mozambique Channel) and other tropical environments: a new addition to the *Planothidium delicatulum* complex. *Fottea, 18*(1), 106–119.

Riaux-Gobin, C., Witkowski, A., Jordan, R.W., Parravincini, V. & Planes, S. (2018d). *Cocconeis kurakakea*, a new diatom species from Nukutavake (Tuamotu Archipelago, South Pacific): description and comparison with *C. diruptoides* and *C. pseudodiruptoides*. *Phytotaxa, 349*(2), 115–129.

Riaux-Gobin, C., Guerrero, J. M., Ector, L., Witkowski, A., Blanco, S. & Daniszewska-Kowalczyk, G. (2019a). *Cocconeis carinata* sp. nov. (Bacillariophyceae) and re-examination of *Cocconeis orbicularis* Frenguelli & H.A.Orlando and *Cocconeis reticulata* var. *deceptionis* Frenguelli & H.A.Orlando. *Diatom Research, 34*(3), 149–163. https://doi.org/10.1080/0269249x.2019.1646321

Riaux-Gobin, C., Igersheim, A., Ector, L., Banaigs, B. & Witkowski, A. (2019b). *Cocconeis scutellum* var. *ornata* Grunow and *C. interrupta* Grunow from the historical Grunow collection in Vienna (sample 131). *Phytotaxa, 408*(1), 41–58. https://doi.org/10.11646/phytotaxa.408.1.3

Riaux-Gobin, C., Witkowski, A., Bemiasa, J. & Bemanaja, E. (2019c). *Cocconeis nosybetiana* sp. nov. from Nosy Be Island (Madagascar) and allied taxa. *Nova Hedwigia, 108*(3–4), 321–338.

Riaux-Gobin, C., Ashworth, M.P., Kociolek, J.P., Chevallier, D., Saenz-Agudelo, P., Witkowski, A., Daniszewska-Kowalczyk, G., Gaspar, C., Lagant, M., Touron, M., Carpentier, A., Stabile, V. & Planes, S. (2021a). Epizoic diatoms on sea turtles and their relationship to host species, behaviour and biogeography: a morphological approach. *European Journal of Phycology, 56*(4), 359–372. https://doi.org/10.1080/09670262.2020.1843077

Riaux-Gobin, C., Frankovich, T., Witkowski, A., Saenz-Agudelo, Esteve, P., Ector, L. & Bemiasa, J. (2021b). *Cocconeis tsara* sp. nov., *C. santandrea* sp. nov. and allied taxa pertaining to the new section Loculatae. *Phytotaxa, 484*(2), 145–169.

Riaux-Gobin, C., Garcia, M., Witkowski, A., Saenz-Agudelo, P., Coste, M. & Daniszewska-Kowalczyk, G. (2021c). New *Amphicocconeis* (Bacillariophyta) from Raivavae and Tahiti Islands (South Pacific) and Porto Belo (Brazil), with re-examination of *Psammococconeis*. *Phytotaxa, 513*(1), 30–54. https://doi.org/10.11646/phytotaxa.513.1.2

Riaux-Gobin, C., Saenz-Agudelo, P., Górecka, E., Witkowski, A., Daniszewska-Kowalczyk, G. & Ector, L. (2021d). *Cocconeis vaiamanuensis* sp. nov. (Bacillariophyceae) from Raivavae (South Pacific) and allied taxa: ultrastructural specificities and remarks about the polyphyletic genus *Cocconeis* Ehrenberg. *Marine Biodiversity, 51*, 29. https://doi.org/10.1007/s12526-020-01154-9

Riaux-Gobin, C., Witkowski, A., Risjani, Y., Yunianta, Y., Berteaux-Lecellier, V., Kryk, A., Peszek, Ł. & Daniszewska-Kowalczyk, G. (2022a). *Upsilococconeis dapalistriata* gen. nov. & comb. nov. (Bacillariophyta), a pantropical marine member of the Cocconeidaceae. *Oceanological and Hydrobiological Studies, 51*(1), 23–31. https://doi.org/10.26881/oahs-2022.1.03

Riaux-Gobin, C., Witkowski, A., Coste, M., Berteaux-Lecellier, V. & Daniszewska-Kowalczyk, G. (2022b). Marine Achnanthales (Bacillariophyta) from New Caledonia (Melanesia): assemblage specificities, ultramafic environment. *Phytotaxa, 552*(1), 1–21. https://doi.org/10.11646/phytotaxa.552.1.1

Ricard, M. (1975). Quelques diatomées nouvelles de Tahiti décrites en microscopie photonique et électronique à balayage. *Bulletin du Muséum National d'Histoire Naturelle, Série 3, Botanique, 326*(3), 202–229.

Ricard, M. (1977). Les peuplements de diatomées des lagons de l'archipel de la Société (Polynésie française): Floristique, écologie, structures des peuplements et contributions à la production primaire. *Revue algologique nouvelle série, 12*(3–4), 138–336.

Rivera, F., Vasselon, V., Jacquet, S., Bouchez, A., Ariztegui, D., Rimet, F. (2018). Metabarcoding of lake benthic diatoms: from structure assemblages to ecological assessment. *Hydrobiologia, 807*, 37–51. https://doi.org/10.1007/s10750-017-3381-2

Romero, O. & Jahn, R. (2013). Typification of *Cocconeis lineata* and *Cocconeis euglypta* (Bacillariophyta). *Diatom Research, 28*(2), 175–184.

Romero, O.E. & Navarro, J.N. (1999). Two marine species of *Cocconeis* Ehrenberg (Bacillariophyceae): *C. pseudomarginata* Gregory and *C. caribensis* sp. nov. *Botanica Marina, 42*, 581–592.

Rougerie, F., Wauthy, B. & Rancher, J. (1992). Le récif barrière ennoyé des îles Marquises et l'effet d'île par endo-upwelling. *Comptes rendus de l'Académie des sciences, 315*, série 2, 677–682.

Rougerie, F., Fichez, R. & Déjardin, P. (1997). Geomorphology and hydrogeology of selected islands of french Polynesia: Tikehau (atoll) and Tahiti (barrier reef). In: H.L. Vacher & T. Quinn (eds), *Geology and Hydrogeology of Carbonate Islands. Developments in Sedimentology, 54*, 475–502.

Round, F.E. & Bukhtiyarova, L. (1996). Four new genera based on *Achnanthes (Achnanthidium)* together with a re-definition of *Achnanthidium*. *Diatom Research, 11*(2), 345–361.

Sabbe, K., Vanhoutte, K., Lowe, R., Bergey, E., Biggs, B., Francoeur, S., Hodgson, D. & Vyverman, W. (2001). Six new *Actinella* (Bacillariophyta) species from Papua New Guinea, Australia and New Zealand: Further evidence for widespread diatom endemism in the Australasian region. *European Journal of Phycology, 36*(4), 321–340. https://doi.org/10.1017/S0967026201003328

Schmidt, A., Schmitz, M., Fricke, F., Müller, O., Heiden, H. & Hustedt, F. (1874–1959). *Atlas der Diatomaceen-Kunde.* 4 Bände, pls. 1–480. Leipzig.

Seddon, A.W.R., Witkowski, A., Froyd, C.A., KurzydłOwski, K.J., Grzonka, J. & Willis, K.J. (2014). Diatoms from isolated islands II: *Pseudostaurosira diablarum*, a new species from a mangrove ecosystem in the Galápagos Islands. *Diatom Research, 29*(2), 201–211, https://doi.org/10.1080/0269249X.2013.877084

Silva, P.C. (1962). Classification of algae. In R.A. Lewin (ed.), *Physiology and biochemistry of algae* (pp. 827–837). Academic Press, New York and London.

Siqueiros Beltrones, D.A., López-Fuerte, F.O., Martínez, Y.J. & Altamirano-Cerecedo, MdC. (2021). A First Estimate of Species Diversity for Benthic Diatom Assemblages from the Revillagigedo Archipelago, Mexico. *Diversity, 13*(10), 1–47. https://doi.org/10.3390/d13100458

Sovereign, H.E. (1958). The diatoms of Crater Lake, Oregon. *Transactions of the American Microscopical Society, 77*(2), 96–134.

Stancheva, R. (2018). Cocconeis cascadensis, a new monoraphid diatom from mountain streams in Northern California, USA. *Diatom Research, 33*(4), 471–483.

Sterrenburg, F.A.S. (1987). *Anorthoneis*, een assepoester genus? *Diatomededelingen, 2*(1), 11–16.

Stidolph, S.R., Sterrenburg, F.A.S., Smith, K.E.L. & Kraberg, A. (2012). *Stuart R. Stidolph Diatom Atlas*. U.S. Geological Survey Open-File Report 2012–1163, available online at https://pubs.usgs.gov/of/2012/1163/

Suzuki, H. & Tanaka, J. & Nagumo, T. (2001). Morphology of the marine diatom *Cocconeis pseudomarginata* Gregory var. *intermedia* Grunow. *Diatom Research, 16*, 93–102. https://doi.org/10.1080/0269249X.2001.9705511

Suzuki, H. & Kobayashi, A. (2002). Epiphytic diatoms on a red alga *Neorhodomela aculeata* (Prestenko) Masuda from Hokkaido, Japan. *Bulletin of Aoyama Gakuin Senior High School, 25*, 62–77.

Suzuki, H., Nagumo, T. & Tanaka, J. (2005). A new marine diatom *Cocconeis sagaraensis* Hid. Suzuki (Bacillariophyceae) from Japan. *The Journal of Japanese Botany, 80*, 176–185.

Tuji, A. (2009). Examination of type materials and typification of seven diatoms described by C.G. Ehrenberg. In Y. Tanimura & Y. Aita (eds.), Joint Haeckel and Ehrenberg Project: Reexamination of the Haeckel and Ehrenberg Microfossil Collections as a Historical and Scientific Legacy. *National Museum of Nature and Science Monographs, 40*, 13–21.

Van Heurck, H. (1880). *Synopsis des Diatomées de Belgique* Atlas. pls I-XXX [pls 1–30]. Anvers: Ducaju et Cie.
Venn, J.M.A. (1880). On the diagrammatic and mechanical representation of propositions and reasonings. *The London, Edinburgh, and Dublin Philosophical Magazine and Journal of Science, 10*(59), 1–18. https://doi.org/10.1080/14786448008626877
Verleyen, E., Van de Vijver, B., Tytgat, B., Pinseel, E., Hodgson, D.A., Kopalová, K., Chown, S.L., Van Ranst, E., Imura, S., Kudoh, S., Van Nieuwenhuyze, W., Sabbe, K. & Vyverman, W. (2021). Diatoms define a novel freshwater biogeography of the Antarctic. *Ecography, 44*, 548–560. https://doi.org/10.1111/ecog.05374
Vyverman W., E. Verleyen, A. Wilmotte, D. Hodgson, A. Willems, K. Peeters, B. Van de Vijver, A. De Wever, F. Leliaert & K. Sabbe (2010). Evidence for widespread endemism among Antarctic micro-organisms. *Polar Science, 4*(2), 103–113.
Witkowski, A. (1993). *Cocconeis hauniensis* sp. nov., a new epipsammic diatom from Puck Bay (southern Balstic Sea), Poland. *Nordic Journal of Botany, 13*(4), 467–471.
Witkowski, A., Lange-Bertalot, H. & Metzeltin, D. (2000). Diatom flora of marine coasts I. *Iconographia Diatomologica, 7*, 1–925.
Zimmermann, J., Glöckner, G., Jahn, R., Enke, N. & Gemeinholzer, B. (2015). Metabarcoding vs. morphological identification to assess diatom diversity in environmental studies. *Molecular Ecology Resources, 15*(3), 526–542. https://doi.org/10.1111/1755-0998.12336
Zupo, V., Jüttner, F., Maibam, C., Butera, E. & Blom, J.F. (2014). Apoptogenic metabolites in fractions of the benthic diatom *Cocconeis scutellum parva*. *Marine Drugs, 12*(1), 547–567. https://doi.org/10.3390/md12010547

Figures 4–30

Fig. 4. *Amphicocconeis antiqua* Riaux-Gobin & Coste. SEM. SV with one row of cupules on each side of the sternum (a, b arrows), SVVC with short fimbriae (d), RV densely striated, composed of a short and strongly bent macroareola near the raphe and a longer one up to margin (c), RVVC extended, open, simple pores up to margin (e, f). Figs 3b, e unpublished. Scale bars: 6 µm (e), 5 µm (a), 4 µm (c), 3 µm (d), 2 µm (b, f).

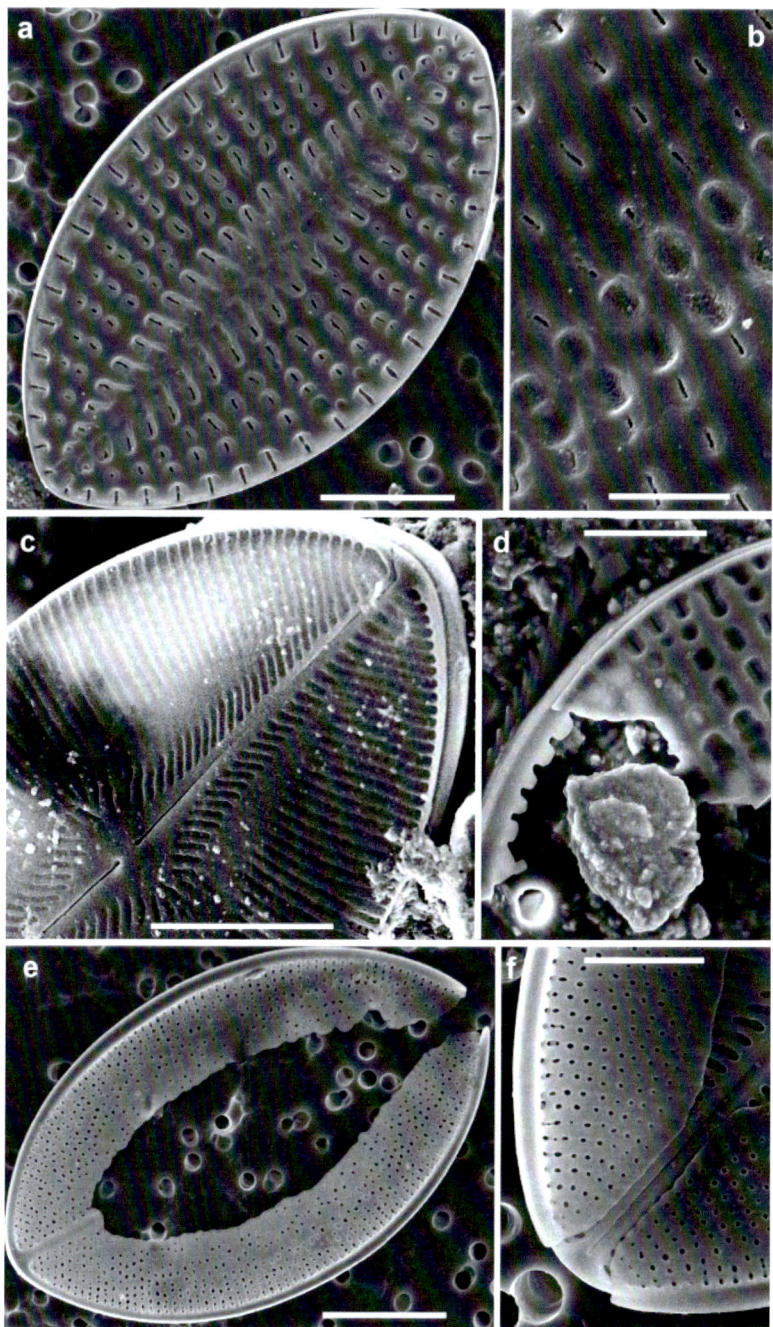

Fig. 5. *Amphicocconeis clypeus* Riaux-Gobin & Witkowski. SEM. SV with one marginal row of denser areolae (a), open large SVVC with fused fimbriae and elliptical void areas-pores (b), RV in external view, with a row of short macroareola near the raphe and longer ones on the margin (c), RVVC extended, open, with small pores up to margin (d). Figs 5a, b, d unpublished. Scale bars: 5 µm (c), 4 µm (a, b, d).

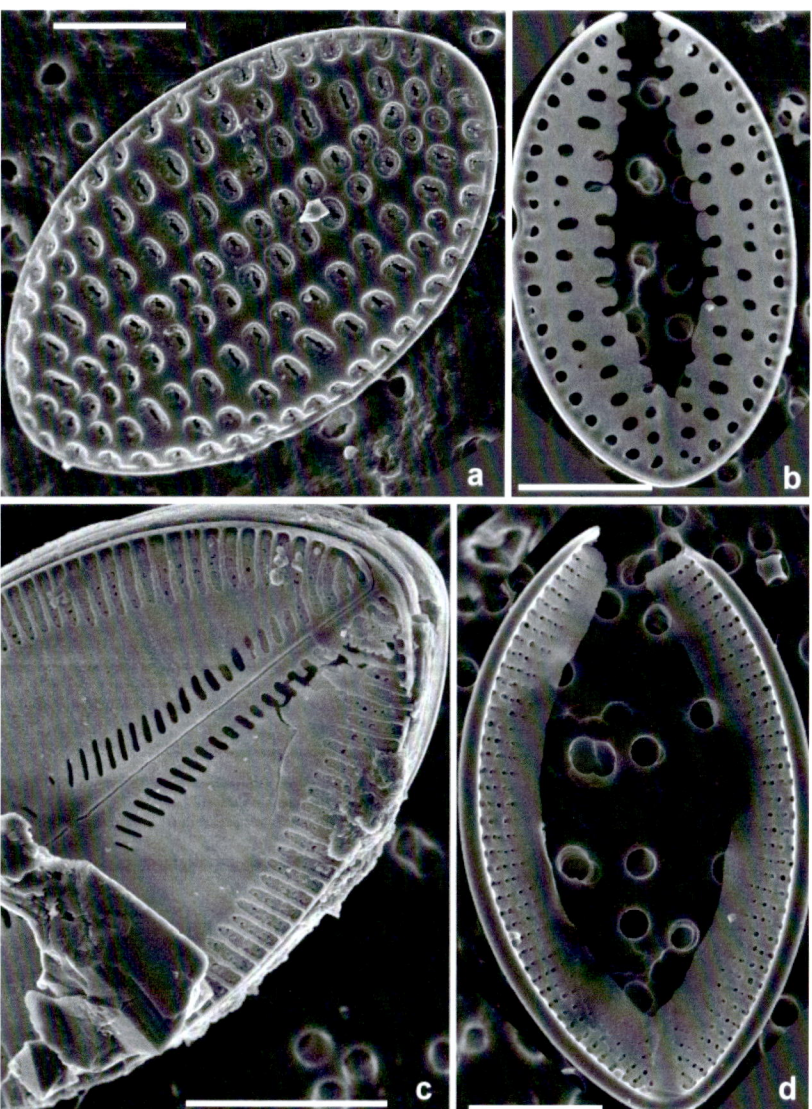

Fig. 6. *Amphicocconeis ruatara* Riaux-Gobin. SEM. Oblong-elliptic to elongate SV with one marginal row of dense and short areolae (external view, a; internal view, b), detail of the SV areolae in external view (d), open narrow SVVC composed of short and fused fimbriae (c, e, f). Figs 6a–g unpublished. Scale bars: 3 µm (a, b, e), 1 µm (c, d), 700 nm (f).

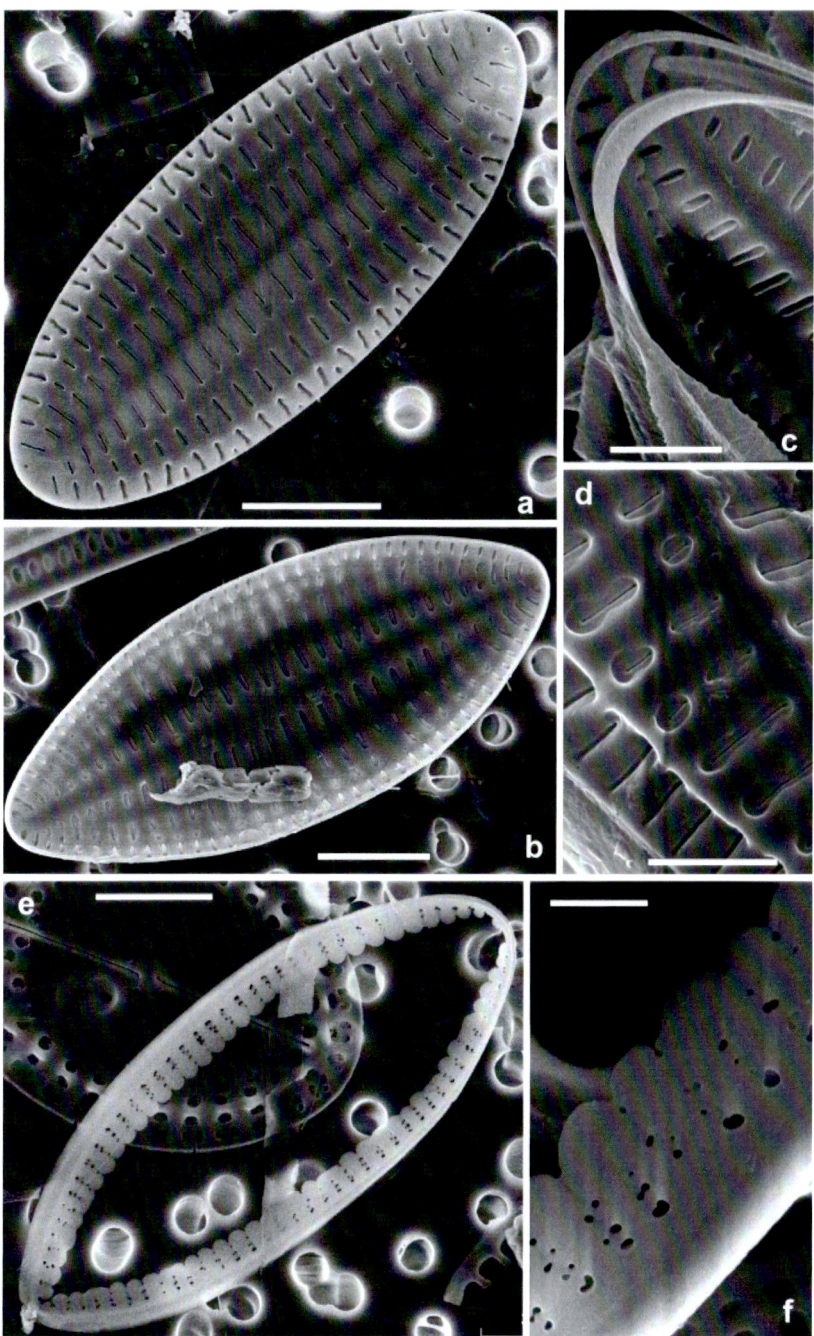

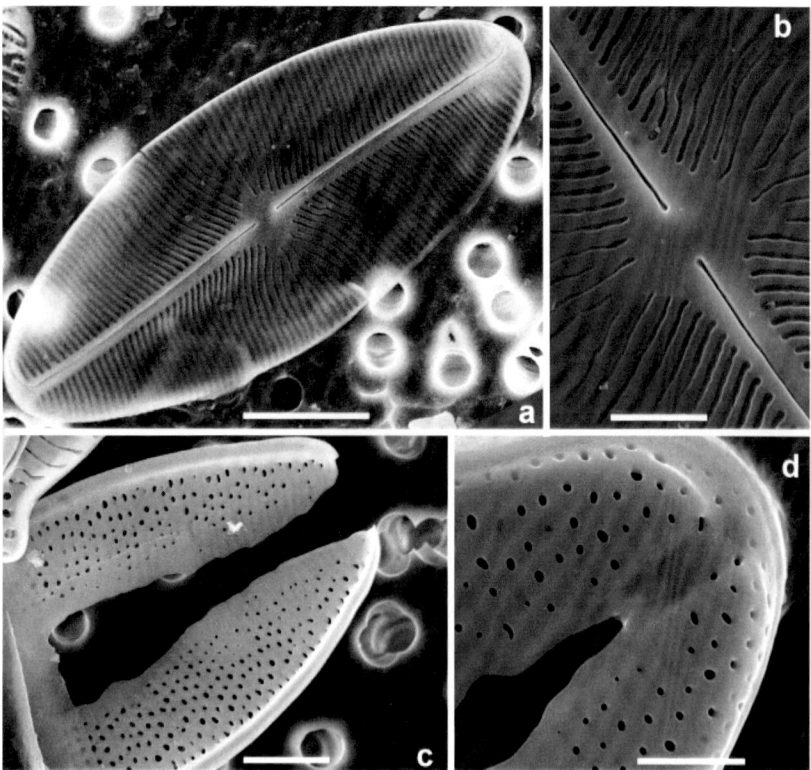

Fig. 7. *Amphicocconeis ruatara* Riaux-Gobin. SEM. RV flat, dense striae composed of one narrow macro-areola (external view, a), central area small (external view, b), RVVC open, extended, with rows of uniseriate pores (c,d). Figs 7c, d unpublished. Scale bars: 3 μm (a), 2 μm (c), 1 μm (b, d).

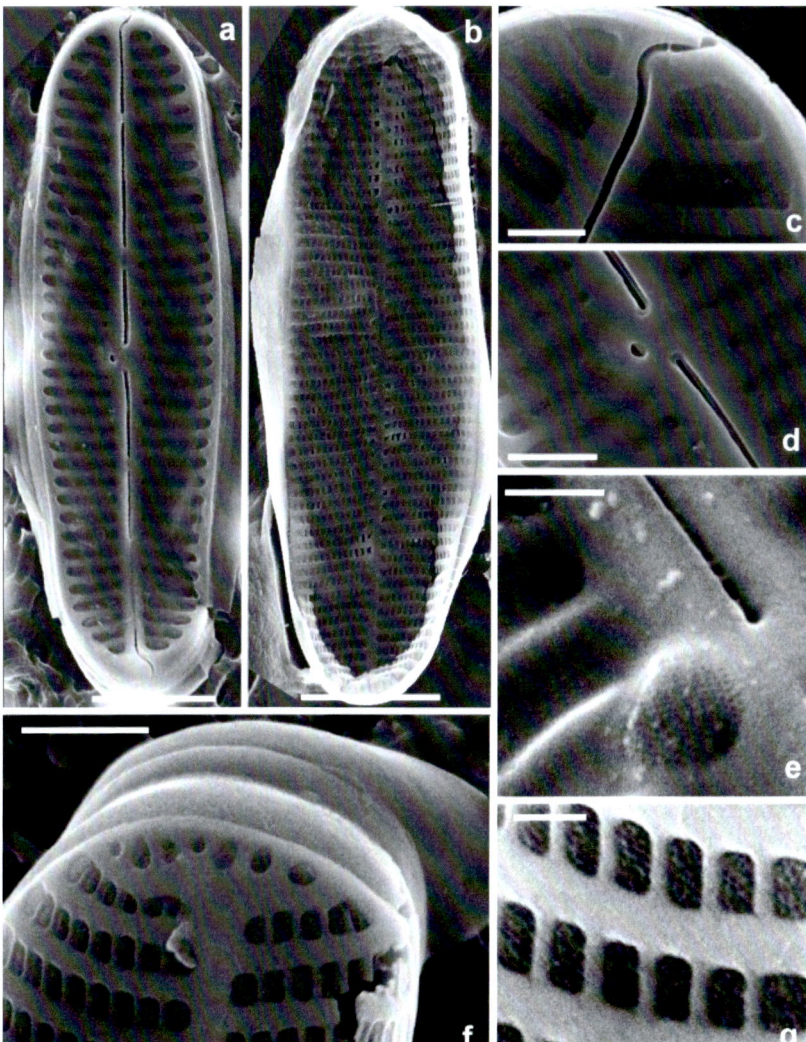

Fig. 8. *Astartiella societatis* Riaux-Gobin, Witkowski & Romero. SEM. RV in external view (a), detail of the doubly hooked raphe ending in external side (c), RV central area with the opening of the stigma (or fistula) (d), internal view of the system with crystal-like appearance (e), SV apex in external view, with one row of small areolae (f), detail of SV areolae, with granular hymen (g). Figs 8a, b, d, f, g unpublished. Scale bars: 2 µm (a, b), 600 nm (d), 400 nm (f), 300 nm (c), 200 nm (e), 100 nm (g).

Fig. 9. *Cocconeis frustrationis* Riaux-Gobin, Compère & Jordan. SEM. SV in external view with a particular marginal row of biseriate areolae (a, b), SV areola hymen in external view, with dots and no slits, closed SVVC in abvalvar side (d), SVVC margin on advalvar side (e), internal view of SV areolae with granular surface (f). Figs 9d unpublished. Scale bars: 2 µm (a, b, d), 1 µm (e, f), 400 nm (f), 300 nm (c).

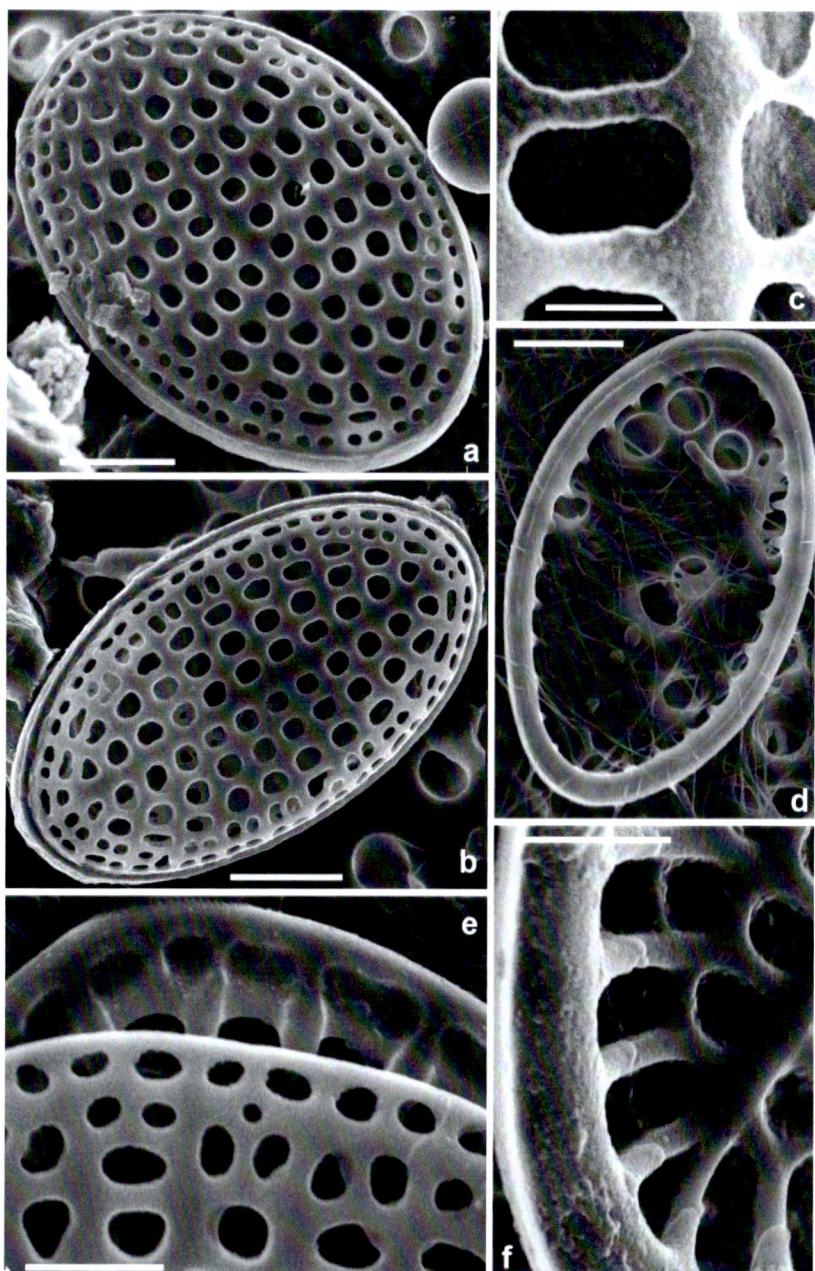

Fig. 10. *Cocconeis frustrationis* Riaux-Gobin, Compère & Jordan. SEM. RV in external view (a), detail of the RV apex in external view (b), closed RVVC with stylet-like fimbriae (c), detailed in (d). Figs 10a–d unpublished. Scale bars: 3 µm (c), 2 µm (a), 1 µm (b), 700 nm (d).

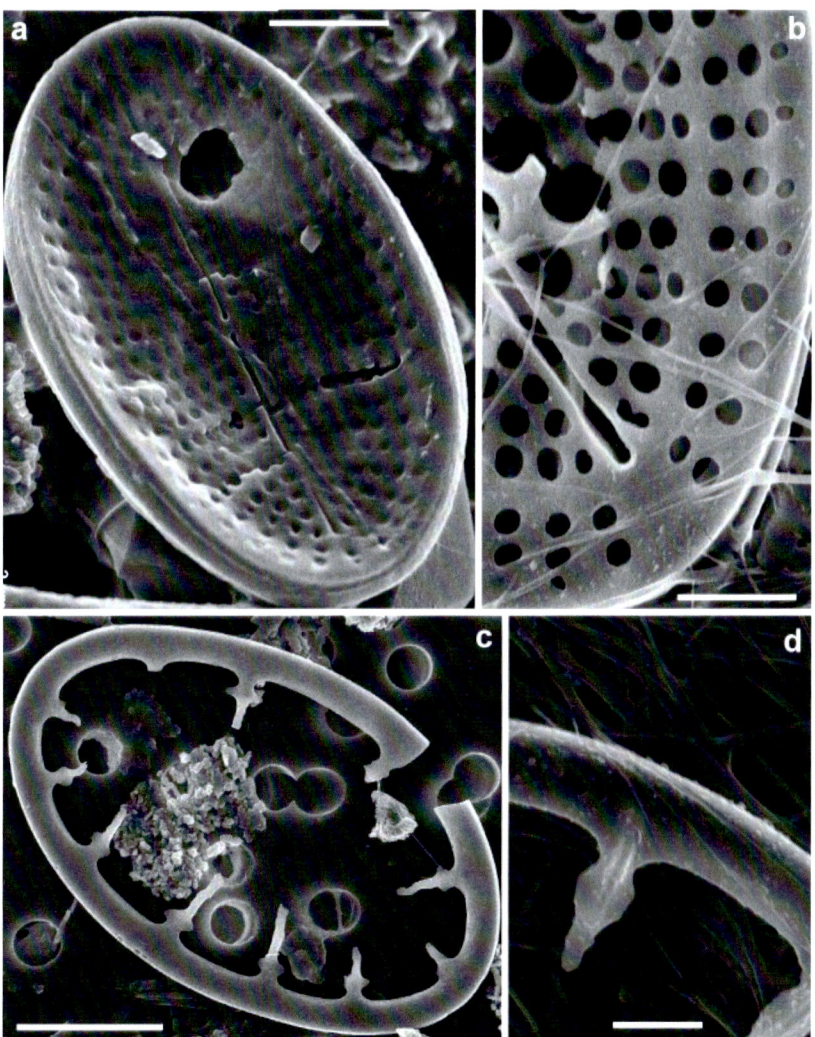

Fig. 11. *Cocconeis kurakakea* Riaux-Gobin & Witkowski. SEM. SV in external view with specific orientation of apical striae (a), SV internal view of the valve apex with SVVC with no fimbriae (b), SV in internal view (d), RV in external view of a large specimen (c) with RV narrow fascia, RV apex in internal view, with prominent helictoglossa and RVVC with no fimbriae. Figs 11a, c, d unpublished. Scale bars: 2 µm (a, c, d), 2 µm (c), 1 µm (e), 600 nm (b).

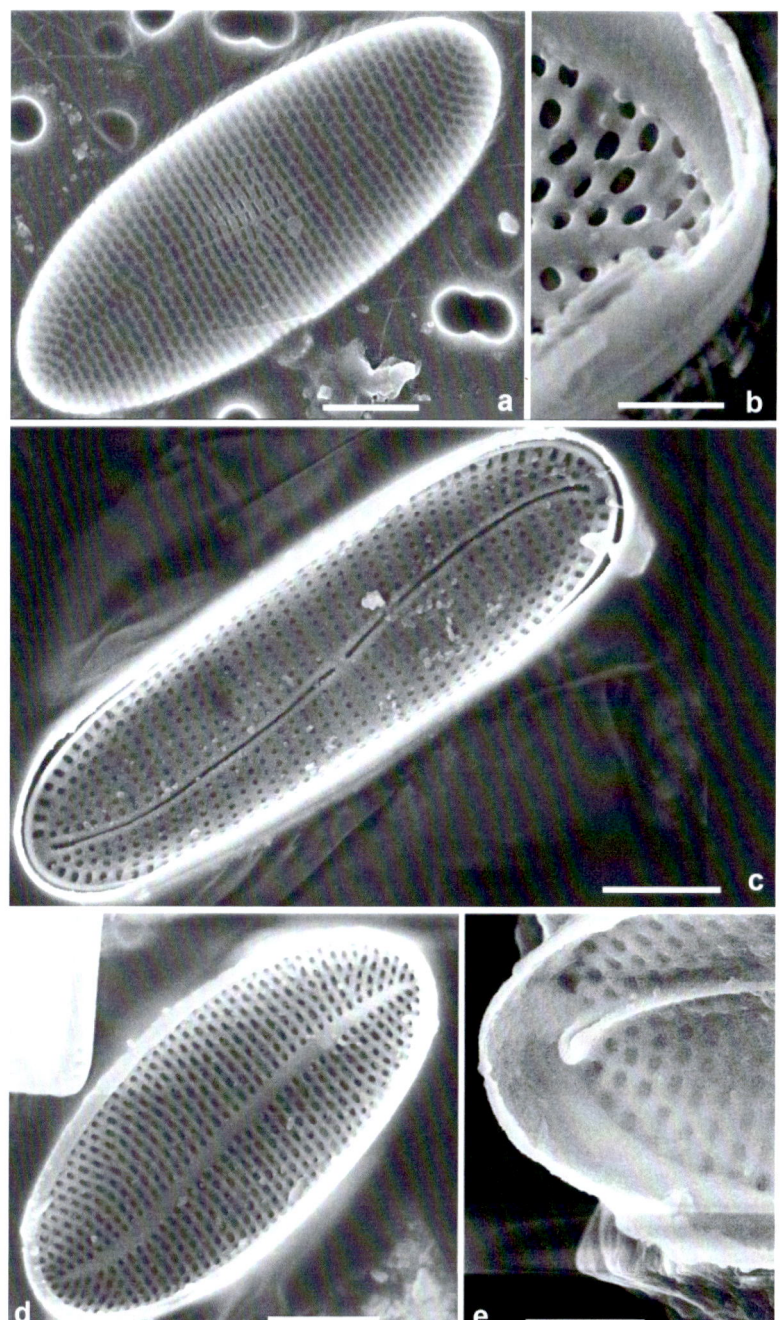

Fig. 12. *Cocconeis santandrea* Riaux-Gobin, Witkowski & Bemiasa. SEM. SV in external view with characteristic SV areolae with San Andrew cross (a) detailed in (c), RV in external view (b), marginal biseriate RV striae with alternate areolae (in ears of wheat) (d), RV margin in internal view (e) with bumps on the elevated rim. Figs 12a, c unpublished. Scale bars: 5 µm (a, b), 1 µm (d, e), 300 nm (c).

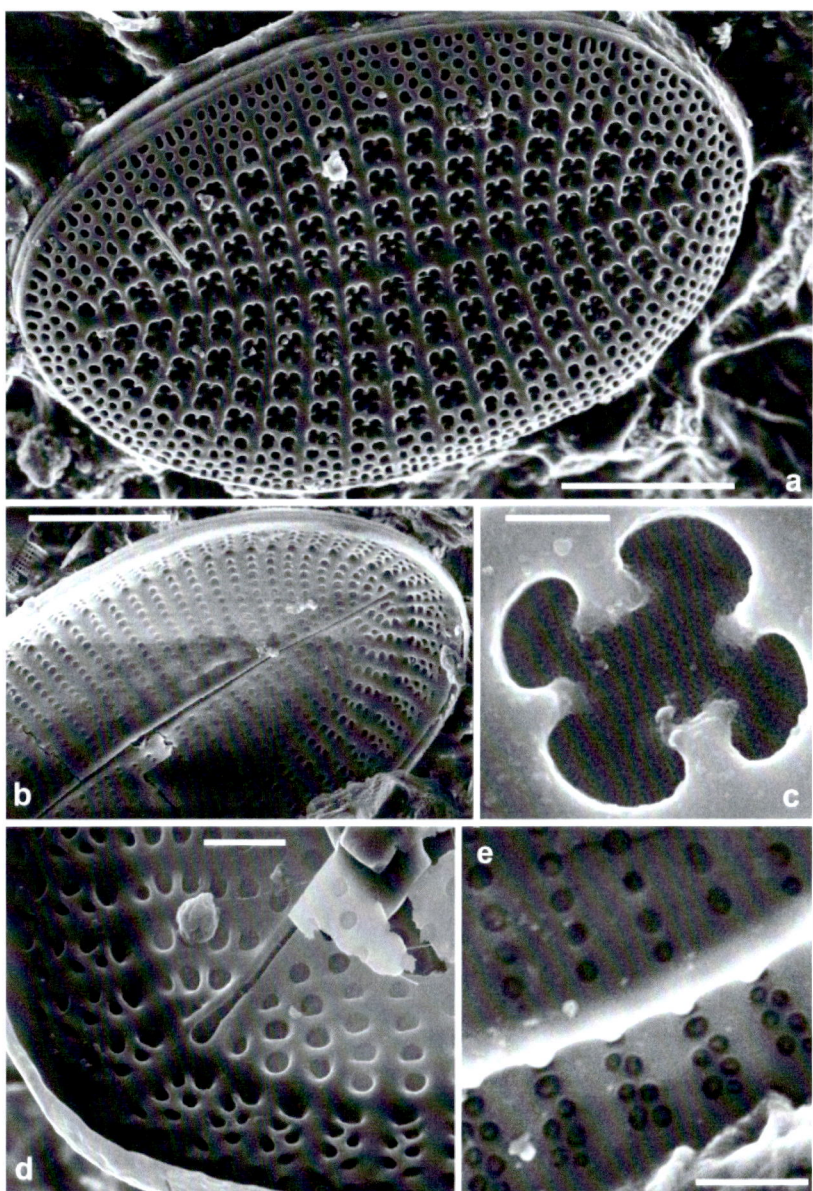

Fig. 13. *Cocconeis spina-christi* Riaux-Gobin, Romero, Coste & Galzin. SEM. SV in external view (a), with detail of the SV domed hymenes (c), SV in internal view, with SVVC showing cupules and digitate fimbriae (b), detail of the SVVC still in place, with cupules and arrowhead fimbriae (e), SVVC on advalvar side(f), both valvocopulae, detached (d), with the RVVC embossed papillae and large fenestrae (d). Figs 13a–f unpublished. Scale bars: 4 µm (a, b), 3 µm (f), 1 µm (c, d, e).

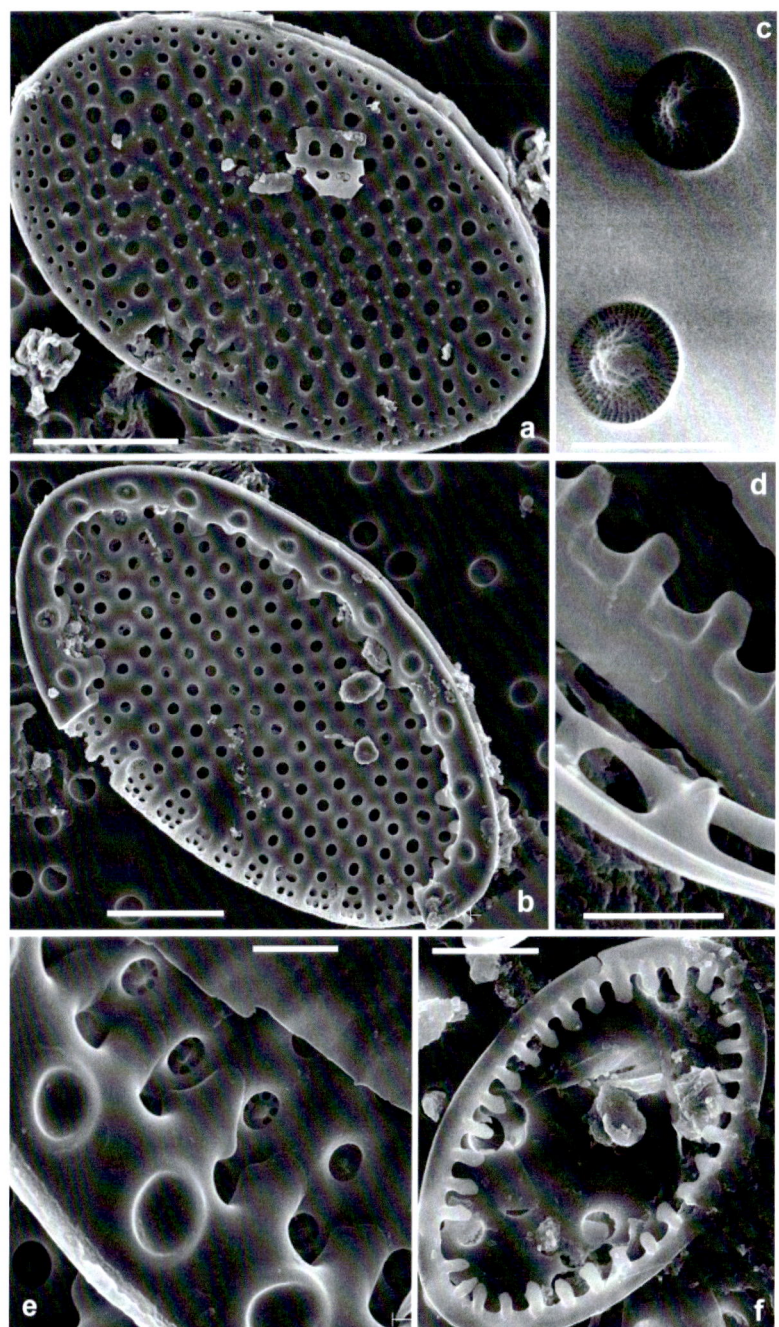

Fig. 14. *Cocconeis spina-christi* Riaux-Gobin, Romero, Coste & Galzin. SEM. RV in external views (a, b, c), with margial RV striation (b), both valvocopulae (c), RVVC in advalvar view, with spines (e), RVVC on abvalvar side, with round papillae (d), with 2 to 3 furrows (f). Figs 14a–d unpublished. Scale bars: 4 µm (a, e), 3 µm (d), 2 µm (c), 1 µm (b), 600 nm (f).

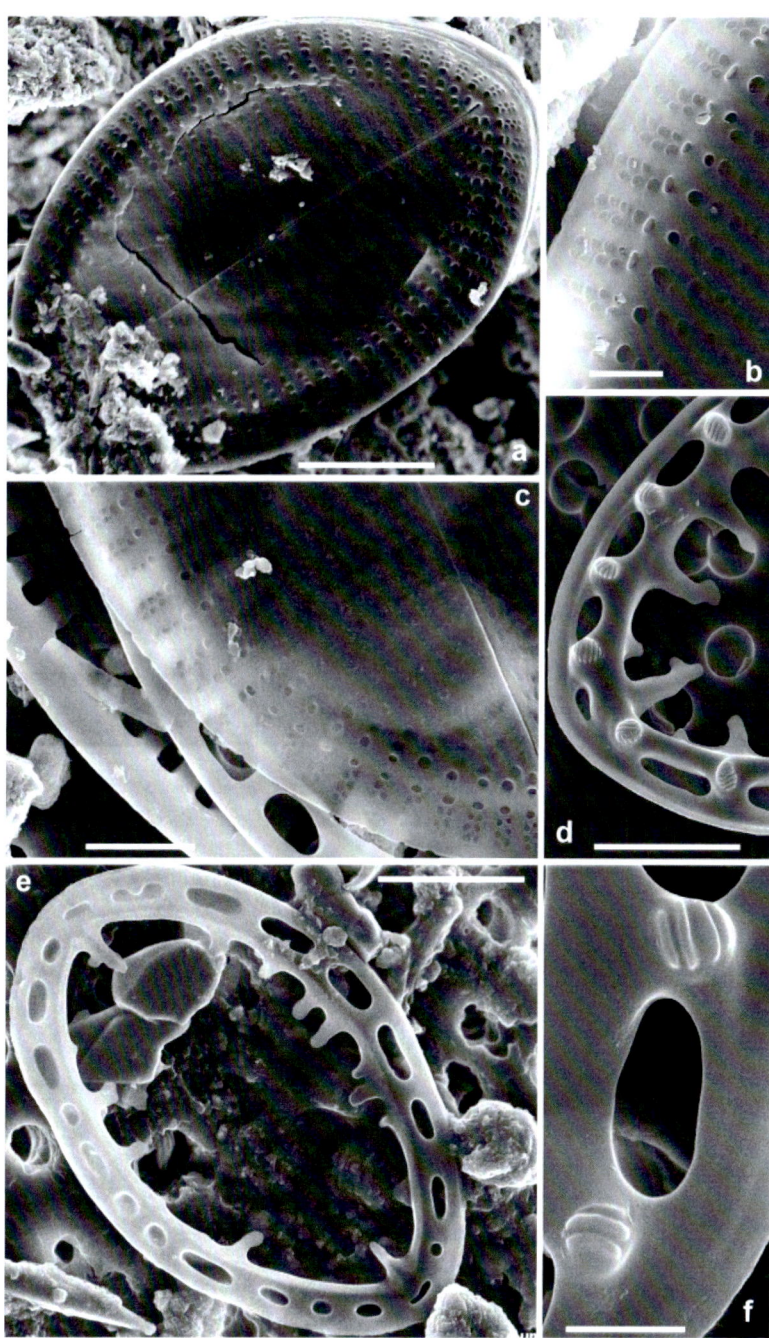

Fig. 15. *Cocconeis tuamotuana* Riaux-Gobin, Compère & Jordan. SEM. SV in external view (a), SV in internal view (c), RV in external view (b), open RVVC (d), RV in internal view with strongly curved helictoglossae (e), RV central area (f). Figs 15a–f unpublished. Scale bars: 5 µm (a, d), 4 µm (b, e), 3 µm (c), 1 µm (f).

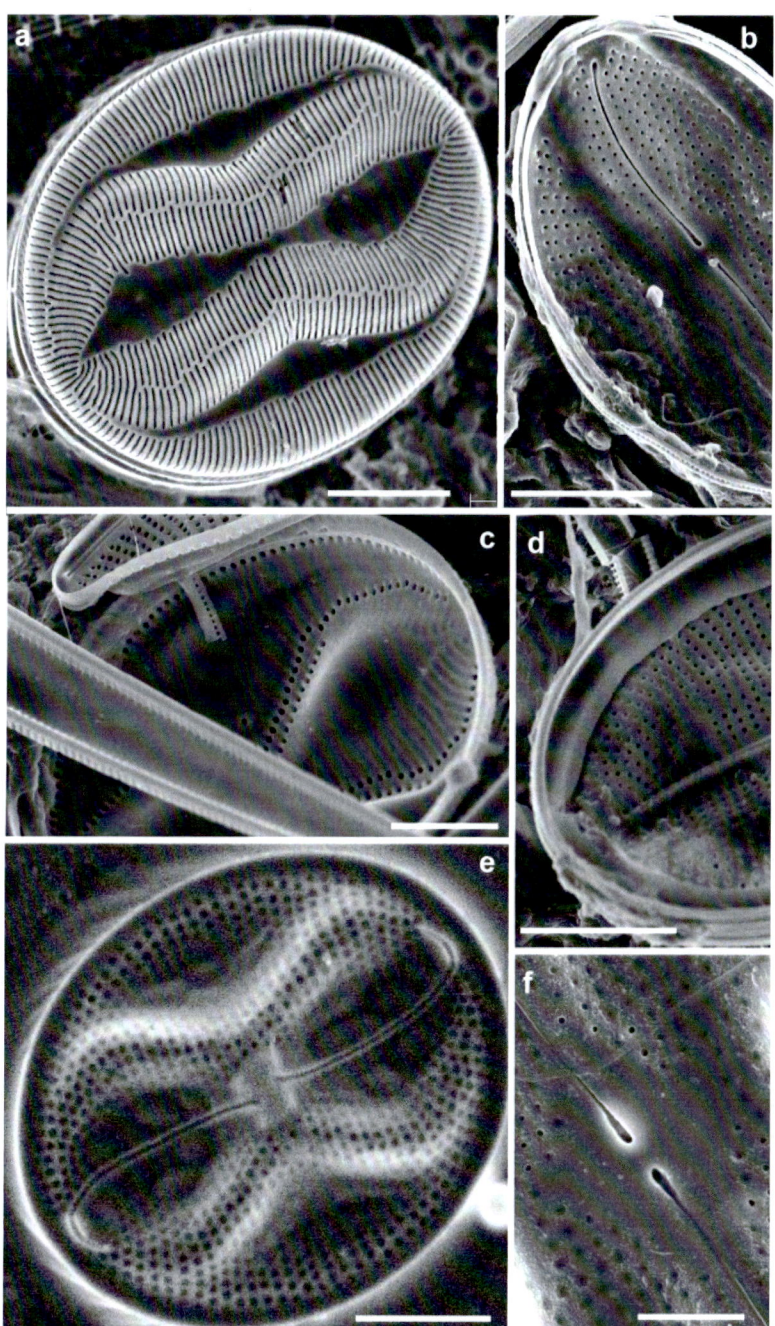

Fig. 16. *Cocconeis vaiamanuensis* Riaux-Gobin, Witkowski & Ector. SEM. SV in external view with marginal simple processes (a), SV in internal view (b), RV in external view with one marginal row of oblong areolae (d), RV areolae in internal view, with marginal simple processes (c), RV central area in internal view (e), raphe terminal ending in internal view (f). Figs 16a–e unpublished. Scale bars: 2 µm (a, b, d), 1 µm (f), 500 nm (e), 300 nm (c).

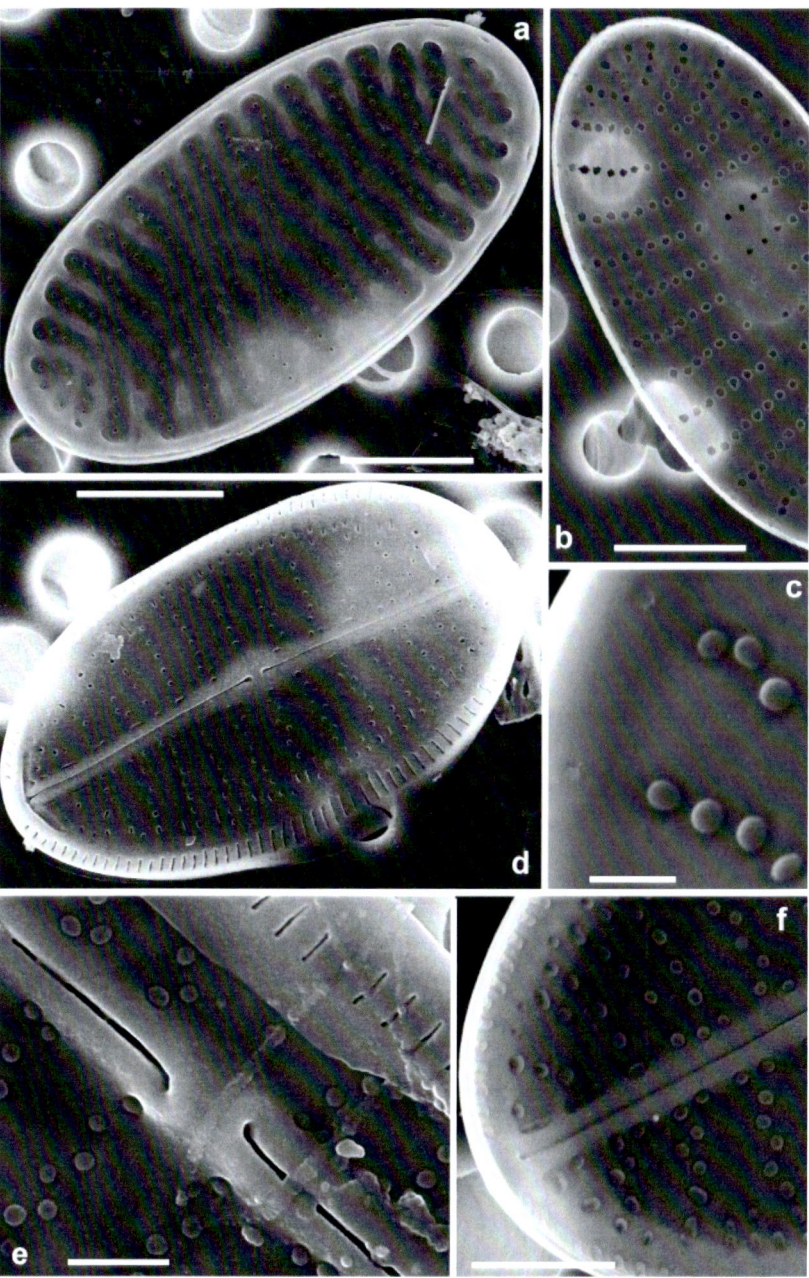

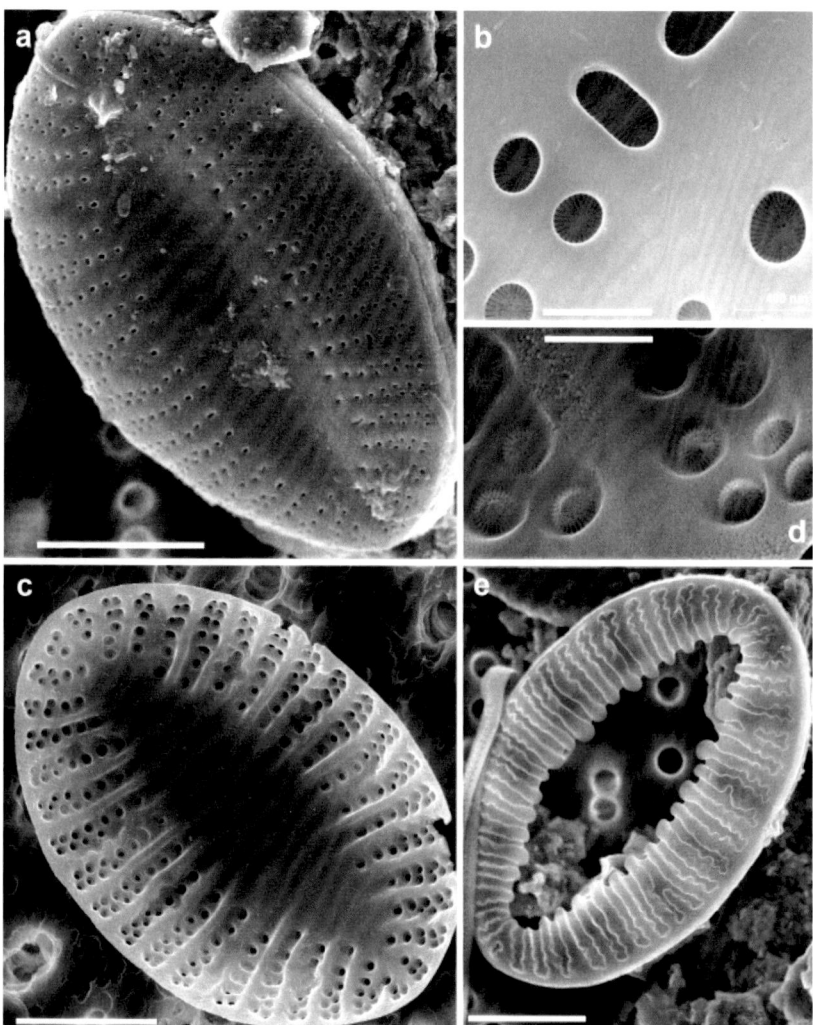

Fig. 17. *Xenococconeis opunohusiensis* Riaux-Gobin, Coste & Romero. SEM. SV in external view (a), with detail of SV areola hymenes (b), and internal view (d), SV in internal view c), SVVC in advalvar view (e). Figs 17a–e unpublished. Scale bars: 4 µm (a, c, e), 400 nm (b, d).

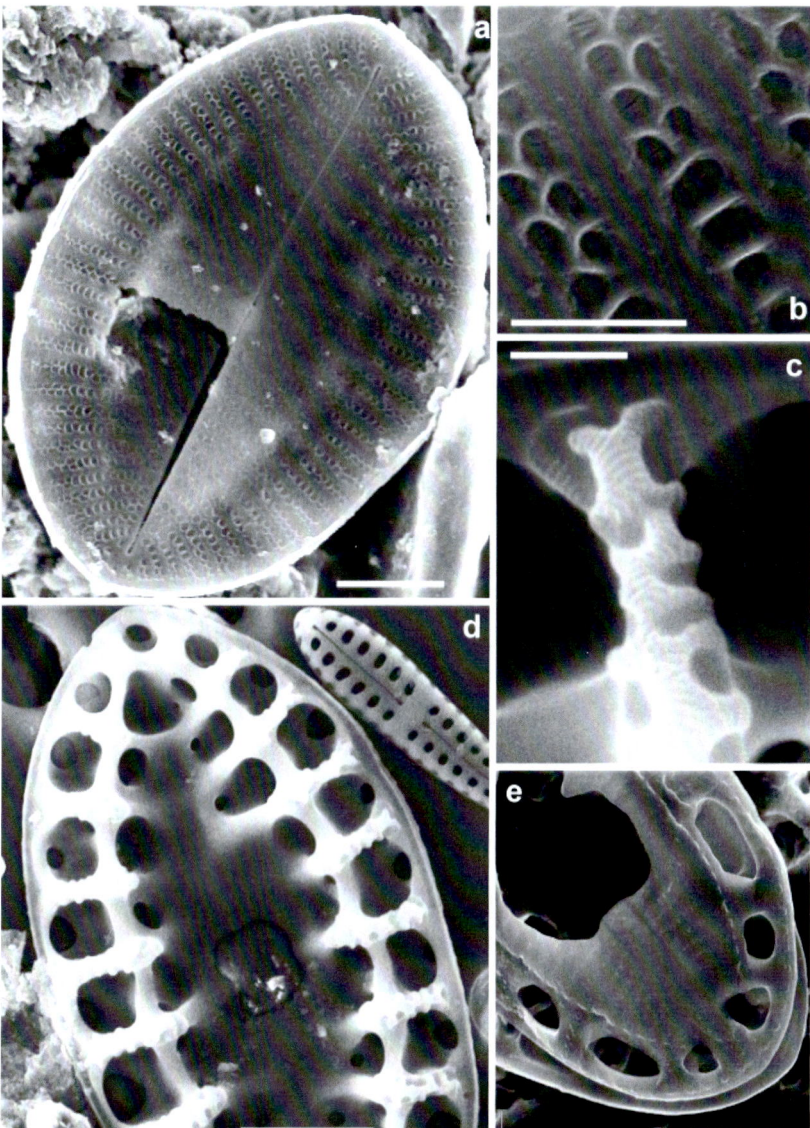

Fig. 18. *Xenococconeis opunohusiensis* Riaux-Gobin, Coste & Romero. SEM. RV in external view (a), with detail of the RV areola hymenes (b), complete RVVC on abvalvar side (d), with detail of the RVVC papillae (c), RVVC on advalvar side with finger-prints of the RV striae, and large marginal holes (e). Figs 18a, b unpublished. Scale bars: 2 µm (a, d, e), 700 nm (c), 500 nm (b).

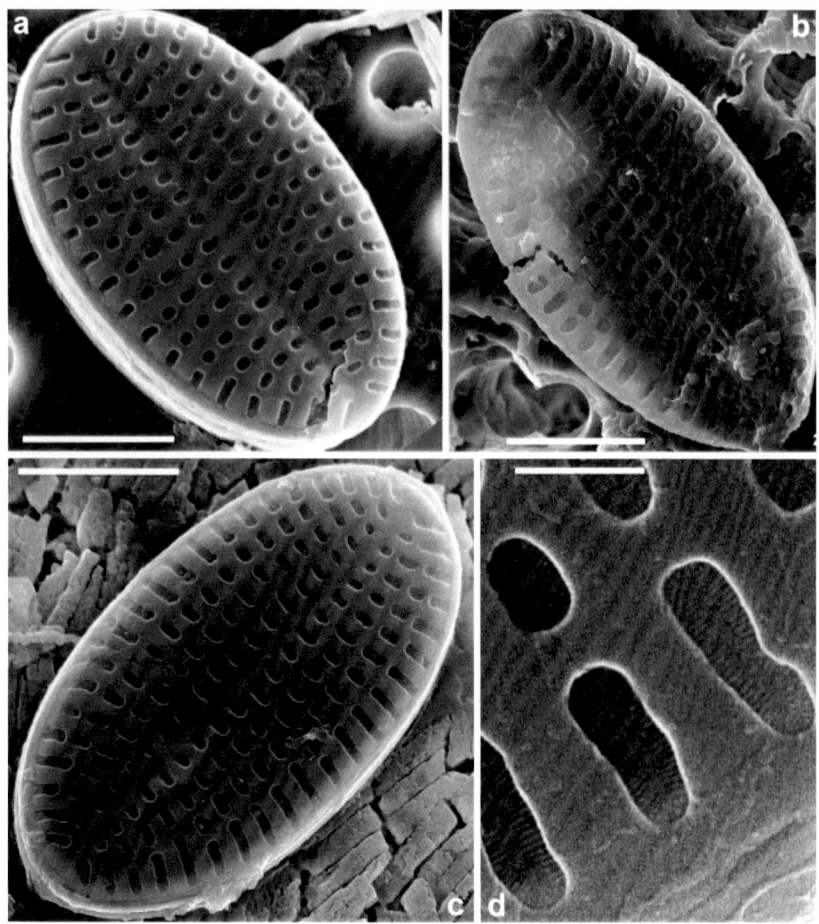

Fig. 19. *Cocconeis angularipunctata* Riaux-Gobin, Romero, Compère & Al-Handal. SEM. SV in external views (a, c), with detail of SV areola hymenes (d), SV in internal view (b). Figs 19a–d unpublished. Scale bars: 2 μm (a, b, c), 300 nm (d).

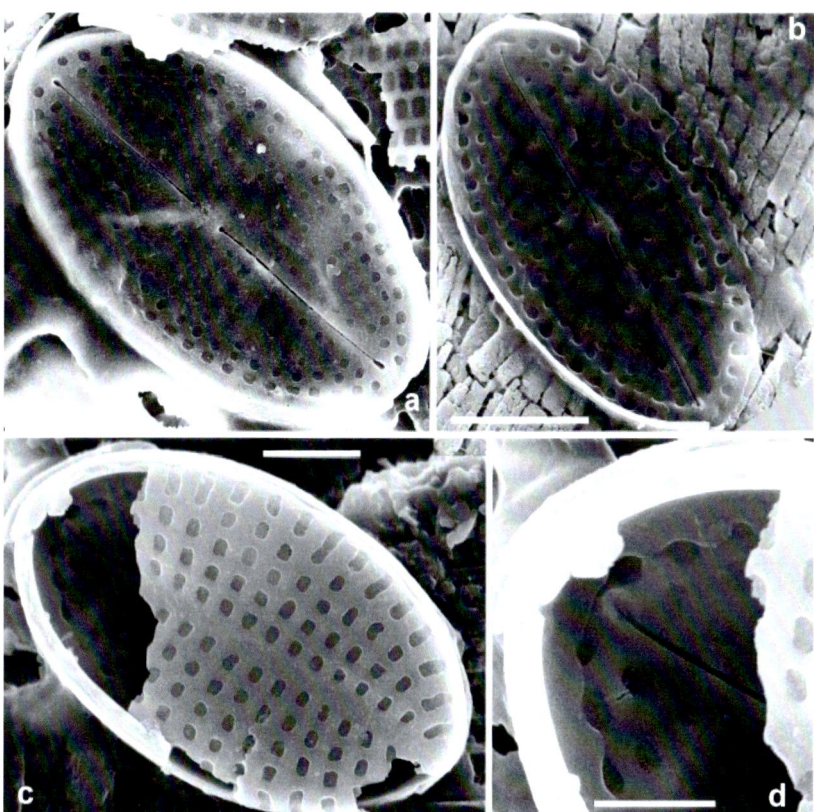

Fig. 20. *Cocconeis angularipunctata* Riaux-Gobin, Romero, Compère & Al-Handal. SEM. RV in external view (a), in internal view (b) with undulated RVVC, both valves (c, d), with detail of the RVVC and helictoglossa (d). Figs 20a–d unpublished. Scale bars: 2 μm (a, b), 1 μm (c), 600 nm (d).

Fig. 21. *Cocconeis geometrica* Riaux-Gobin, Romero, Compère & Al-Handal. SEM. SV in external view (a, b), with detail of SV areola hymenes (b), RV in internal view (c, d), open RVVC, with detail of the irregular RVVC fimbriae (d), RV in external views (e, f). Figs 21a–f unpublished. Scale bars: 2 µm (a, c, e, f), 1 µm (d), 300 nm (b).

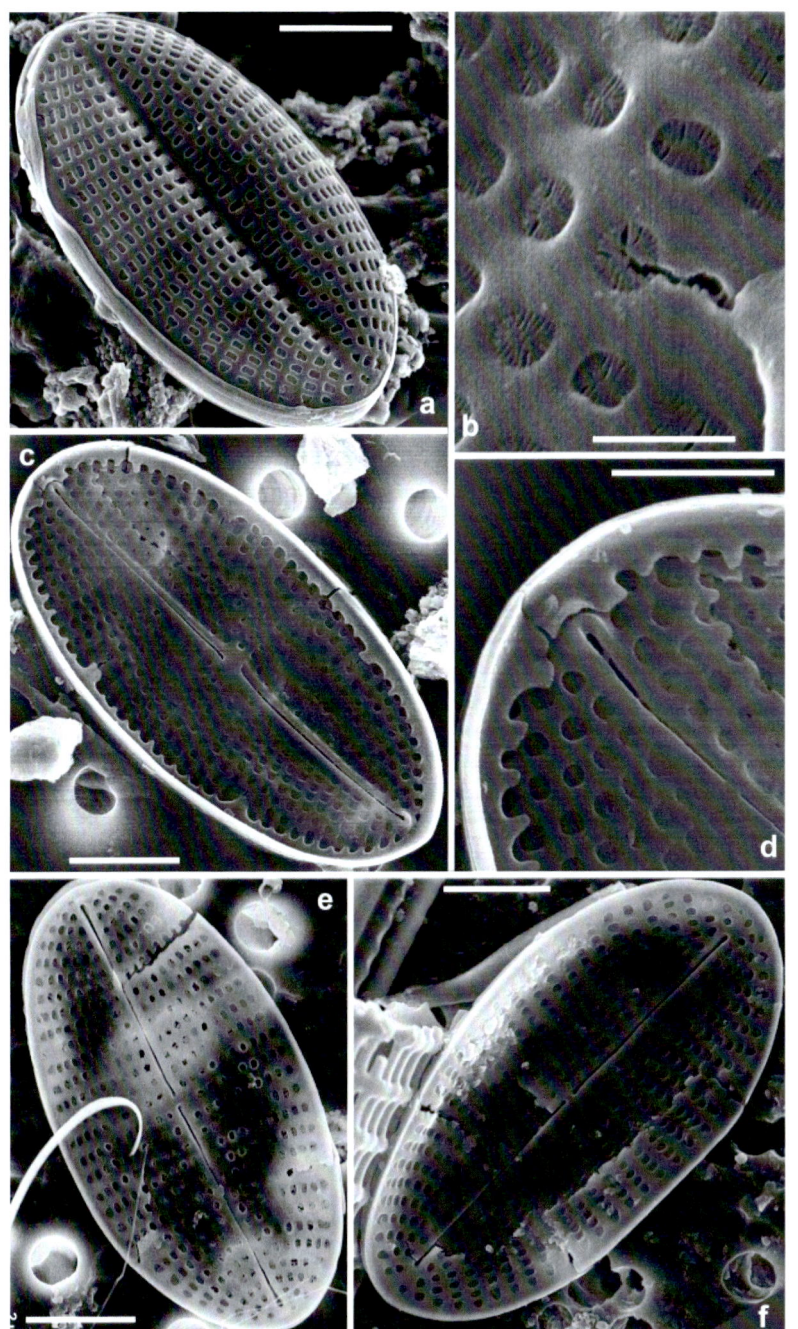

Fig. 22. Rare and unnamed tropical taxa. *Cocconeis* sp. 6 (a–d), *Cocconeis* sp. 7 (e, f). SEM. *Cocconeis* sp. 6 SV in external view (a, b), with detail of the marginal row of simple processes (b), *Cocconeis* sp. 6 SV in internal view (c, d), with detail of simple processes (d), *Cocconeis* sp. 7 SV in external view (e), with marginal simple processes (f). Figs 22e, f unpublished. Scale bars: 3 µm (a), 2 µm (c, e), 1 µm (f), 500 nm (d), 400 nm (b).

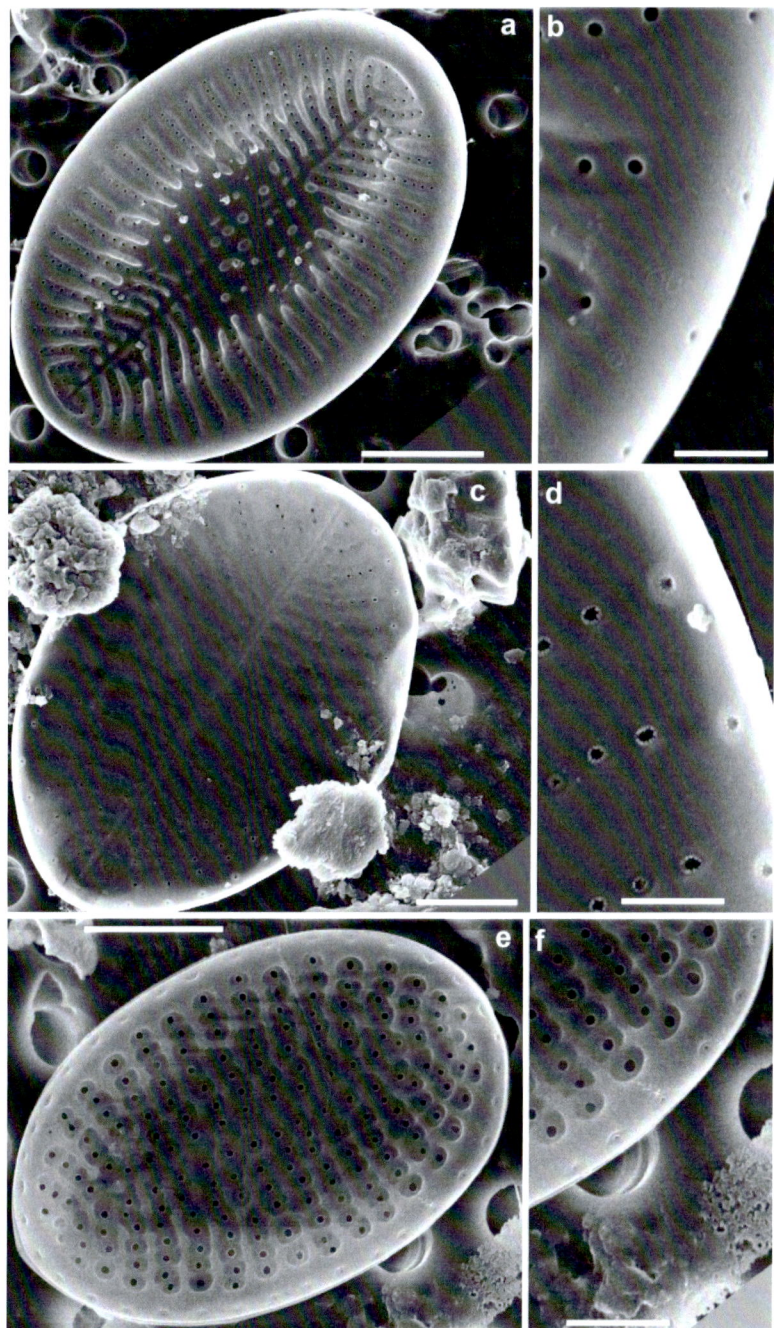

Fig. 23. Rare and unnamed tropical taxa. *Cocconeis* sp. 1. SEM. SV with equidistant striae composed of large round areolae, with beads and broken *crista marginalis* (a), detail of the broken *crista marginalis* (b, arrow), marginal row of biseriate areolae, with a particular arrangement (c, arrow), note beads particularly on the SV sternum (c, arrowhead), detail of SV areolae with slits, and beads (d, arrowhead). Scale bars: 4 µm (a, c), 700 nm (b), 300 nm (d).

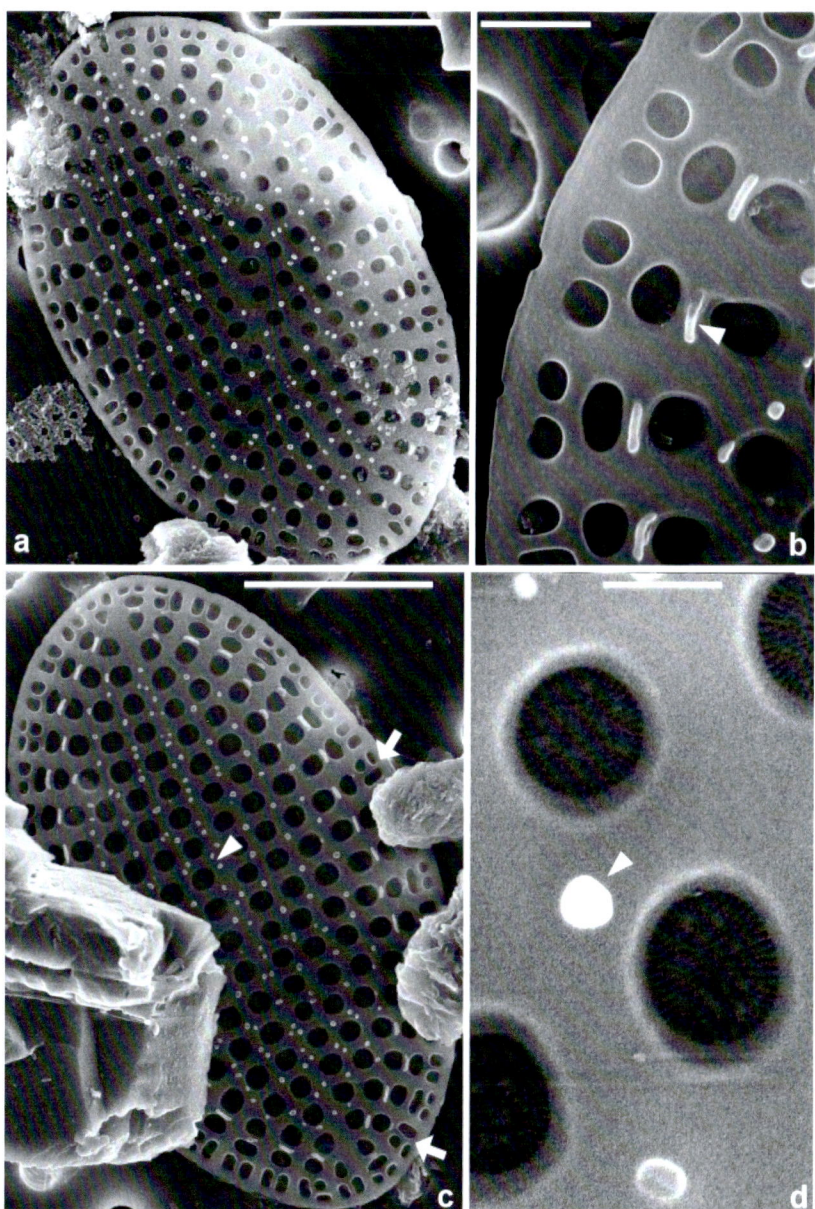

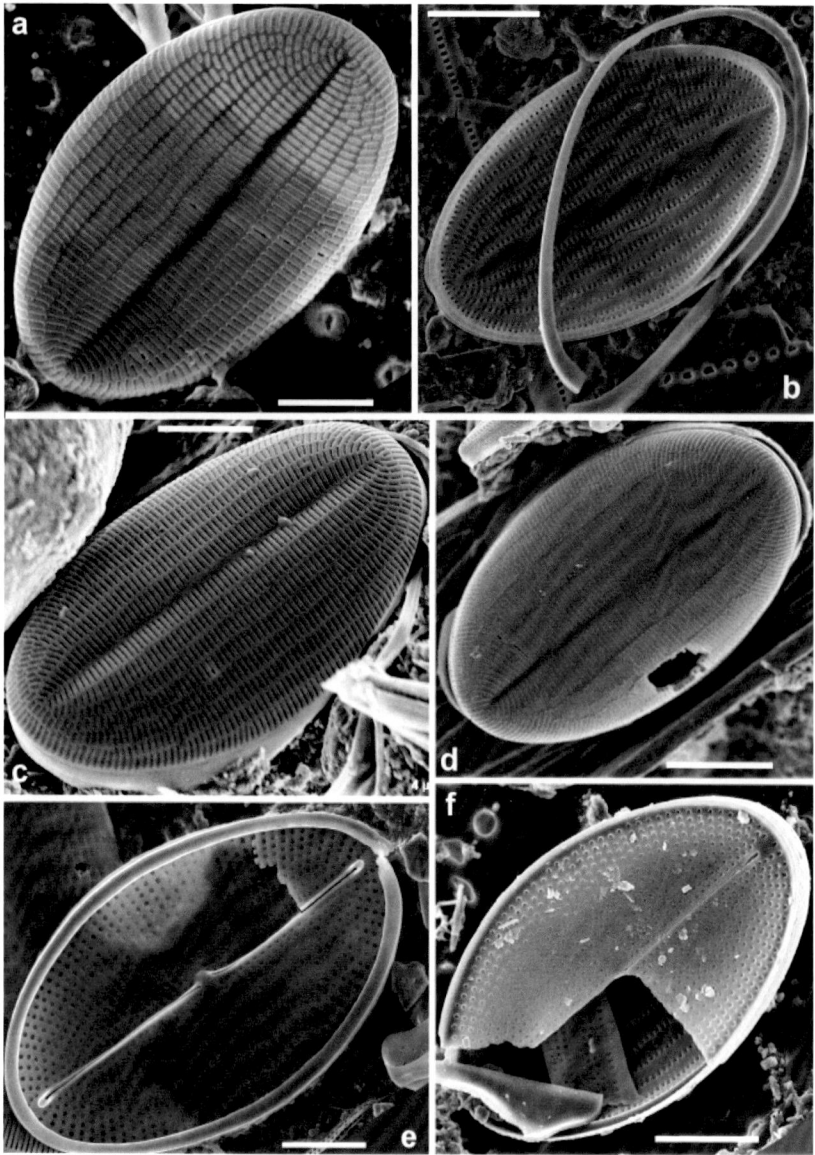

Fig. 24. *Cocconeis convexa* Giffen 'morph'. 'Morph' 1, with oblong valve shape. SEM. SV in external view (a,c), SV in internal view with SVVC open, without fimbriae (b), doubtful specimen with one axial row of longer SV areolae (d), RV in internal view, RVVC without apparent fimbriae, broken RV in external view, with internal view of the SV (f). Scale bars: 5 μm (b, d), 4 μm (c, f), 3 μm (a, e).

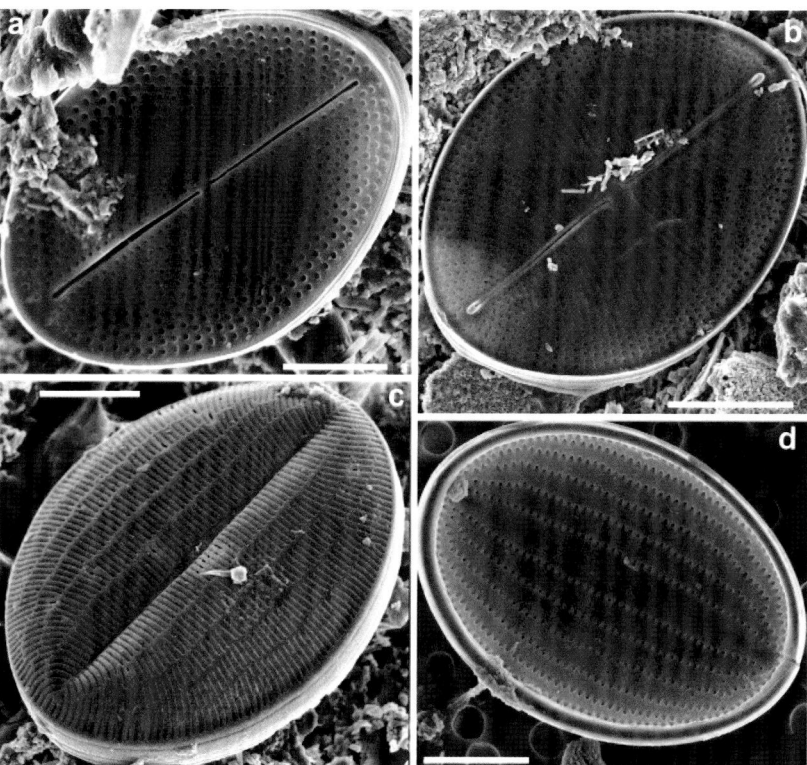

Fig. 25. *Cocconeis convexa* Giffen. Sub-discoid shape (close to original description). SEM. RV in external view (a), RV in internal view (b), SV in external view with axial rows of oblong areolae (c), SV in internal view with rows of pores (giving access to alveoli) (d). Scale bars: 5 µm (b), 4 µm (c), 3 µm (a, d).

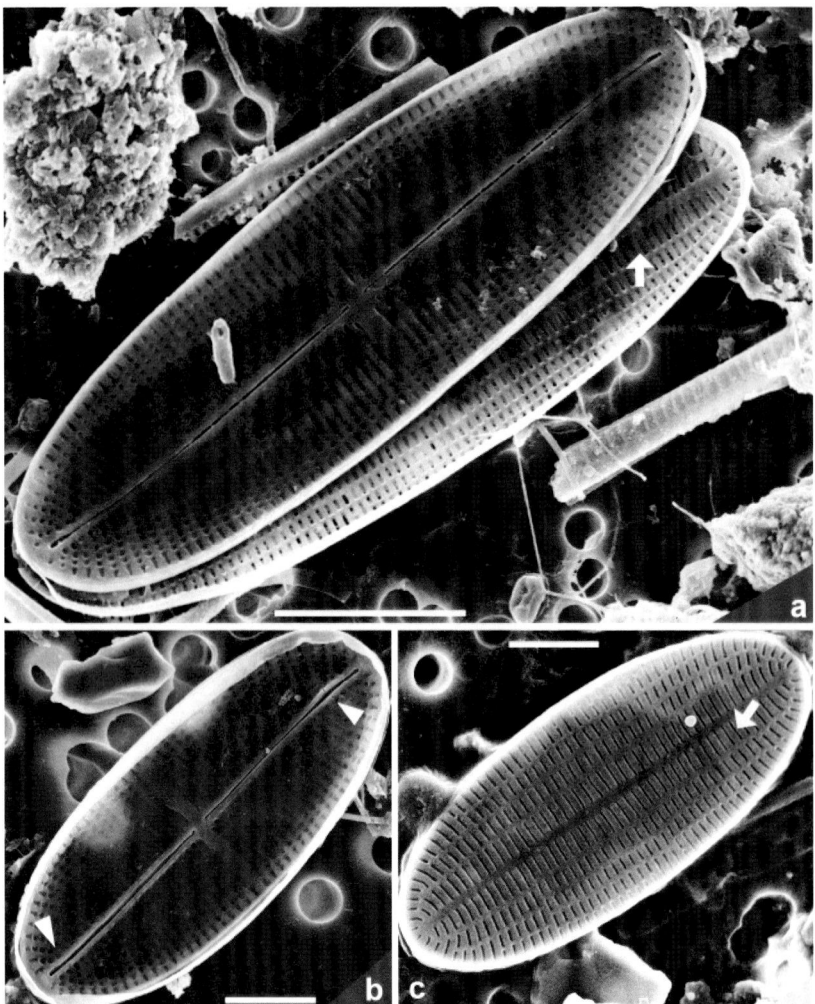

Fig. 26. *Cocconeis borbonica* Riaux-Gobin & Compère. SEM. Both valves of a large specimen, note the row of large areolae on both sides of the SV sternum (arrow) (a), RV on external side (b) note the raphe endings bent in opposite directions (arrowheads), SV on external view (c), with a row of larger areolae on both sides of the SV sternum (c, arrow). Figs 26a–c unpublished. Scale bars: 6 µm (a), 2 µm (b, c).

Fig. 27. *Cocconeis coronatoides* Riaux-Gobin & Romero. 'Morphs' from Polynesia. SEM. Specimens large, subcircular, SV with a *crista marginalis* and siliceous beads (a, d), specimens elongate to linear-elliptic (c, e, f), note the RVVC with irregular thick fimbriae of this 'morph' (f). Figs 27a–f unpublished. Scale bars: 6 µm (b), 5 µm (a), 4 µm (c, e, f), 2 µm (d).

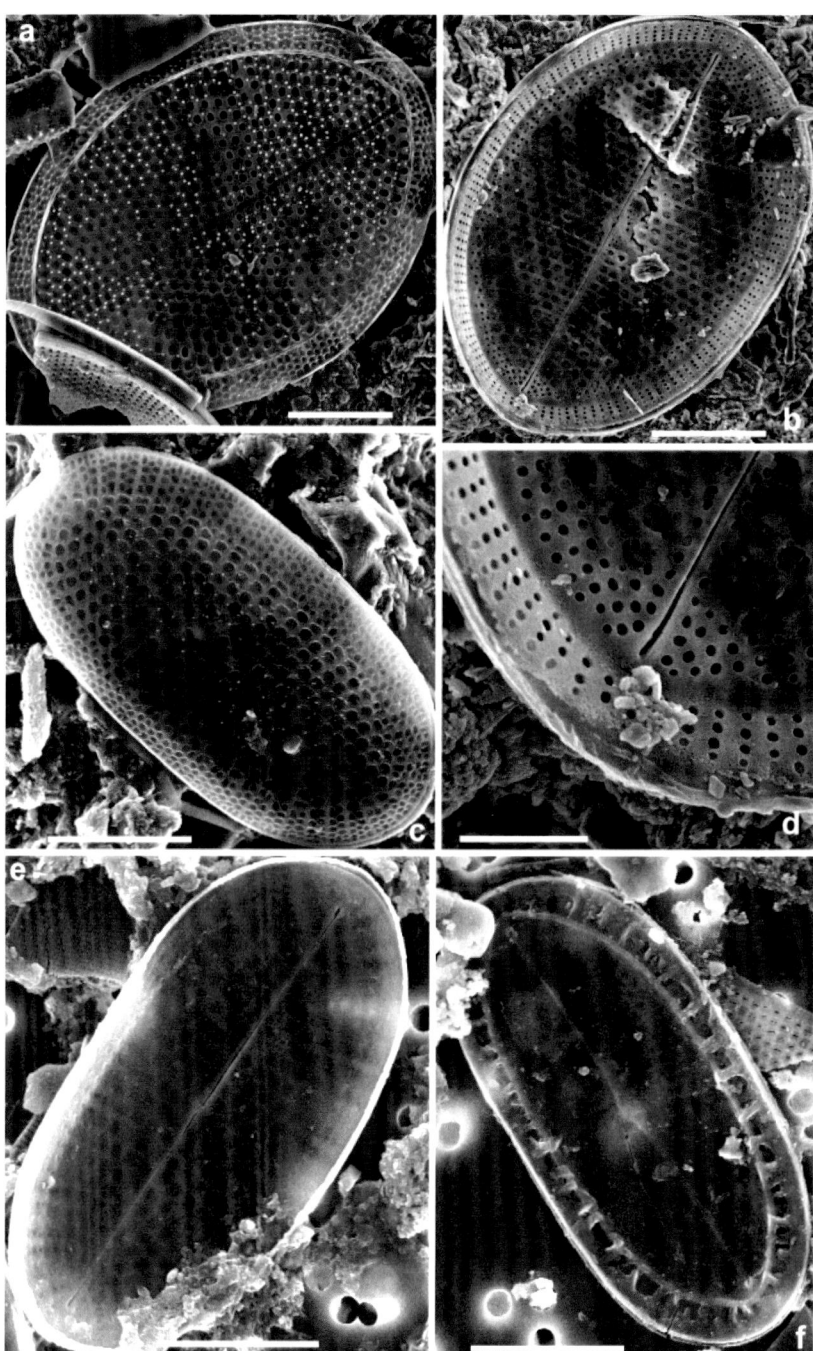

Fig. 28. *Cocconeis placentula* Ehrenberg *sensu lato*. 'Morph' with long SV areolae in axial rows. SEM. SV in external view (a, b), both valves, RVVC with very short undulations (c), detail of the SVVC with very short fimbriae (d), detail of the SV areolae in external view (e). Scale bars: 3 μm (a), 2 μm (b, c), 1 μm (d), 300 nm (e).

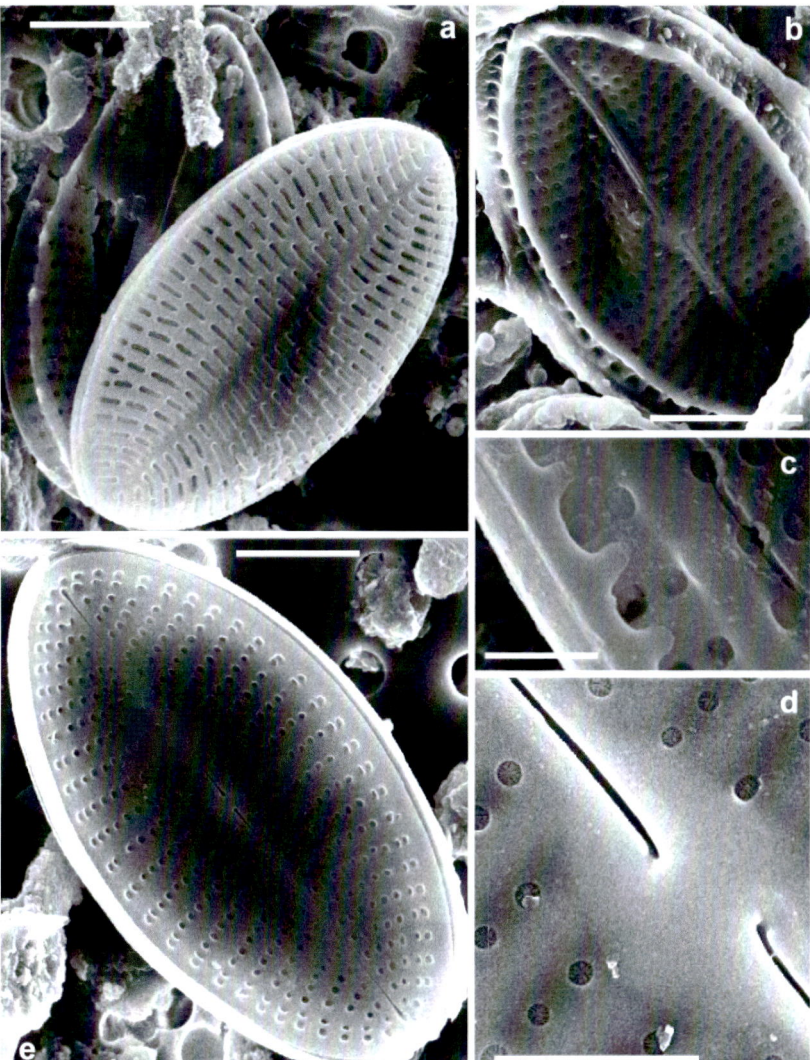

Fig. 29. *Cocconeis placentula* Ehrenberg *sensu lato*. 'Morph' with long SV areolae. SEM. Both valves (a), RV in external view (e), RV in internal view with a marginal rim (b), RVVC with short fimbriae (c), RV central area in external view (d). Scale bars: 3 μm (b), 2 μm (a, e), 700 nm (d), 500 nm (c).

Fig. 30. *Cocconeis placentula* Ehrenberg *sensu lato*. 'Morph' with short SV areolae more or less in quincqunx. SEM. SV external views (a, b, d), RV, internal view with marginal rim (c). Scale bars: 4 μm (b, c), 3 μm (a), 2 μm (d).